SEX LINK

BIRDS DO IT. BEES DO IT.
DO THEY DO IT BETTER
THAN *WE* DO IT?

Here is a fascinating glimpse of the highly diverse, amazingly creative and altogether bizarre mating rituals of the animal world: Music to rape by... master-slave lovers...rampant cuckoldry... bisexual and homosexual love...male chauvinism...group sex...incest...tender passion...fidelity...dance and song—all have their natural place in the mating games of animals. Entertaining, informative and absolutely true, *SEX LINK* may make you realize you have more to learn from "the birds and the bees" than you might like to admit.

SEX LINK

The Three-Billion-Year-Old Urge and What the Animals Do About It

by Hy Freedman

*This low-priced Bantam Book
has been completely reset in a type face
designed for easy reading, and was printed
from new plates. It contains the complete
text of the original hard-cover edition.*
NOT ONE WORD HAS BEEN OMITTED.

SEX LINK

*A Bantam Book / published by arrangement with
M. Evans and Company, Inc.*

PRINTING HISTORY

*M. Evans and Company edition published September 1977
2nd printing . . . September 1977   3rd printing . . . November 1977
Playboy Book Club, October 1977
Psychology Today Book Club, October 1977
Bantam edition / August 1978*

ISBN 0-553-11652-5

Published simultaneously in the United States and Canada

---

*Bantam Books are published by Bantam Books, Inc. Its trade-
mark, consisting of the words "Bantam Books" and the por-
trayal of a bantam, is registered in the United States Patent
Office and in other countries. Marca Registrada. Bantam
Books, Inc., 666 Fifth Avenue, New York, New York 10019.*

PRINTED IN THE UNITED STATES OF AMERICA

0 9 8 7 6 5 4 3 2 1

CONTENTS

1. NATURE DECIDES WHAT IS
   SEXUALLY NORMAL   1

2. IN THE BEGINNING   9

3. HOW DIFFERENT ARE WE
   FROM OTHER ANIMALS?   17

4. SEX—BIZARRE, FANTASTIC,
   BUT COMPLETELY NORMAL   31

   THE EARTHWORM: *Homosexual? Bisexual? Yes*
   THE THREADWORM: *Love Is Where You Find It*
   THE AFRICAN BLOOD FLUKE: *Sexual Menace*
   THE BUMBLEBEE EELWORM:
       The All-Encompassing Vagina
   THE CARP PARASITE WORM: *Permanent
       Love Connection*
   THE STARWORM: *It's Good Work, If You Can Get It*
   THE MARINE BRISTLE WORM: *Castrating Females,
       or Watch Out for That Love Bite*
   THE DEEP-SEA ANGLER: *Her Lover Grows On Her*
   THE GUINEA PIG: *Inventor of the Chastity Belt*
   THE BEDBUG: *No Vagina? Drill Your Own!*
   THE BEDTICK: *Don't Stop Feeding Just Because
       We're Copulating*

5. GROUP SEX   43

   SOCIAL IS SEXUAL
   THE CALLICEBUS MONKEY:
       Love Thy Neighbor—and Then Some
   THE WATER SNAIL: *Hermaphroditic Weirdo*
   THE MUD SNAIL: *Group Orgy in a Chain Gang*
   THE SLIPPER SNAIL: *Togetherness Forever—and Ever*
   THE PARAMECIUM: *A Single Cell with Two
       Ways to Procreate*

THE MAYFLY: *A Lifetime of Hedonism*
THE MOSQUITO: *The Only Way to Fly*
THE BAT: *Doing It Upside Down*
THE FIREFLY: *Orgies in Christmas Trees*

6. RAPE   57

IT TAKES TWO TO TANGLE
THE MINK: *Rapist or Lover?*
THE CROCODILE: *Call the Rape Squad!*
DUCKS AND DRAKES: *Rampant Cuckoldry*
THE FROG: *Music to Rape to*
THE MOTH MITE: *To Rape, Add Incest*
THE PARASITIC WASP: *The Eldest Son Inherits
    All Sexual Privileges*

7. SADOMASOCHISM   73

ONE MAN'S MEAT IS
    ANOTHER MAN'S BOTTOM
THE SNAIL: *Lurid, Sadomasochistic,
    Bisexual Love Life*
BONDAGE AND DISCIPLINE: *Master-Slave Lovers*
THE AFRICAN CLAWED FROG: *How to
    Manhandle a Much Larger Female*
THE LABYRINTH FISH: *Male Chauvinism
    Under Water*
THE RIVER BULLHEAD: *Bullheaded and Bossy*
THE DYSTICID: *A Very Feisty Beetle*

8. SUPERSTUDS AND NYMPHOMANIACS   85

SEX AMONG UNEQUALS
THE LION: *Miserable Male Chauvinism—
    and the Females Love It*
THE FLY: *Super Sexual Athlete*
THE SAGE GROUSE: *A Night to Remember—
    A Year to Recuperate*
THE COCKROACH: *Ugly, but Oh, So Sexy*
THE KOB: *The Owner of the Property
    Gets the Girl*
THE STONE GROUSE: *Beauty Contests—
    Winners and Judges*

9. SEXUAL AIDS   111

THE MANY AVENUES OF SEX
THE OCTOPUS: *Eighteen Arms and
    One Substitute Penis*

THE SPIDER: *Copulate and Run*
THE DRAGONFLY: *A Mating Wheel in Flight*
THE DOG: *Impotency? What's That?*
MASTURBATION: *Even the Animals Do It*
SEX IN THE ZOO: *Captive Sex vs Wild Sex*

10. SEXUAL TENSION   125

SPOUSAL AROUSAL
THE CICHLID: *Don't Play with the Big*
      Fellows Till You're Ready
THE TEN-SPINED STICKLEBACK:
      The Love Bite'll Getcha
THE PORCUPINE: *I'll Let You Know*
      When I'm Ready—and
      Don't Hold Your Breath
THE ROEBUCK: *The Great Chase*
GETTING UPTIGHT FOR
      SUCCESSFUL SEX
THE BEETLE: *The Victor Gets the Spoils*
THE RIVER CRAYFISH
THE ROBIN
THE SALMON

11. THE TENDER RITES OF COURTSHIP   139

THE BOWERBIRD: *Honeymoon in a*
      Real Honeymoon Cottage
THE PHEASANT: *Glad Rags and Happy Sex*
THE RUFF: *Hens Deflate Males*
THE PENGUIN: *Sex on the Rocks*
THE PIGEON: *Open Marriage*
THE RAVEN: *Odd Couples and Triples*
THE GOOSE: *Till Death Do Us Part*
THE HUMMINGBIRD: *Erotic Stunt Flyers*
THE BUTTERFLY: *Beautiful, Fragrant and Sexy*
THE FIDDLER CRAB: *Colorful Beach Boys*
THE TURTLE: *Slow and Easy on the Draw*

12. LOVE AND DEATH   161

THE EEL: *Incredible Navigators*
THE FIREFLY: *Blink Once for Love, Twice for Death*
THE SPIDER: *Love in the Boudoir, or Is It Abattoir?*
THE SCORPION: *The Honeymoon Chamber of Horrors*
THE PRAYING MANTIS: *Please, Not While*
      I'm Copulating
THE SEA URCHIN: *Generation Gap*

13. POPULATION CONTROL 175

THE MACAQUE: *Natural Chastity Belt*
THE LEMMING: *Here Comes the Mob*
THE LOCUST: *Look Who's Coming to Dinner*
THE RAT: *Murder, Rape and Anarchy*
THE ELEPHANT: *Call Me in About
    Five Years, Honey*
THE SPRINGTAIL: *Sex Among the Multitudes*
THE PILL AND OTHER CHEMICALS
SEXUAL ODORS—
    THE KEY TO BIRTH CONTROL

14. FAMILY LIFE 193

THE CHIMPANZEE: *Come and Get It,
    Fellas, I'm in the Pink*
THE GORILLA: *Sexual Score: Feeble Drive,
    Sophisticated Techniques*
THE ORANGUTAN: *The Swingers*
THE BABOON: *It Pays to Advertise*
OTHER PRIMATES: *Polygamy Works*
THE WHALE: *Big Is Everything*
THE SEAL: *No Eating or Sleeping but
    Lots of Loving and Fighting*
THE BEE: *Fifty Thousand Virgins*
THE TERMITE: *Nothing Is Too Good for Royalty*
THE CUCKOO BIRD: *Maybe She's Not So Cuckoo*
THE HORNBILL: *What Confinement Really Means*
THE TALEGALLUS: *Commune Living*
THE FINCH: *The Home Wreckers*
THE CRICKET: *Even the Lonely Need Sex*
THE KANGAROO: *Room Service in a Pouch*
COUVADE: *Male Childbed*
THE VAQUERO FROG: *The Frog Has a Frog
    in His Throat*
THE TOAD: *The Male Midwife*
THE SEA HORSE: *Daddy's Pregnant Again*

INDEX OF ANIMAL LIFE 225

SEX LINK

The Three-Billion-Year-Old Urge
and What the Animals Do About It

1

NATURE DECIDES WHAT IS SEXUALLY NORMAL

The sexual behavior of animals is the theme of this book. Not merely the act of copulation but also the wide range of activities directly or indirectly associated with that physical act are covered here. These include social activities (nurturing of the young, courtship, self-preservation and social intercourse); political activities (yes, there are many animal societies constructed on hierarchical or "pecking order" lines); and economic activities (possession of territory is a fundamental drive among most species).

Then there's the sexual, from the observance of lifelong monogamy to utter promiscuity.

The term "animals" is used in its widest sense, referring to all living organisms aside from plants. This covers everything from the ultramicroscopic virus (it, too, has males, females and a sex life) to the largest mammal that ever lived on earth, the 180-ton blue whale, and about a million different species in between.

If animals could read, and there are infinitely more animals than humans, the distribution of this book would not be affected one whit. If there's one thing that animals don't need more information on, it's sex. That's because sex holds no mystery, is not a vehicle for hang-ups, is never a *cause célèbre* for puritanical censors, and is not a frightening, degrading or unnatural experience to an animal. On the contrary, animals treat sex as a most natural phenomenon, on a par with all the other basic physical activities intrinsic to their everyday lives.

There's good reason for this. The animal is guided by a sure instinct provided by nature, the inventor and chief protagonist of sex. Nature's logic in this respect is impeccable. If her primary function with all living things is to perpetuate the species, and if sex is the only way to accomplish this objective, then sex must be a basic, pleasant and vital necessity to all life. This injunction is observed by all animals on earth, with reservations held by only one species—the modern human. Whether it's because of our larger brains, superior intelligence, tenacious memory, spiritual nature or everpresent conscience with its concomitant offshoots of fear, guilt and anxiety, or some combination of these human qualities, we're certainly having our problems with sex.

This book does not deal with human sexuality except in a tenuous, circuitous way. There's little question that some knowledge of how animals deal with their sexual drives, the sureness, confidence and instinct they display, cannot help but shed some light on the dark nooks and crannies of our own sexuality. After all, these distant cousins share common roots with us; what they're experiencing today is something our ancestors probably experienced in the far distant past.

If we pay close attention to animals' sexual behavior, we may be able to recognize some of our own less orthodox impulses and accept them as part of our physical heritage—and in that way perhaps deal better with our own sexual feelings.

We will read about various species of animals that mate while on the wing, in the water, on land and underground. Once sexually joined, some remain in permanent copulatory union for the duration of their lives; others complete the fertilization process within a few seconds and depart, usually never to meet again. There are species which include within their mating procedures such activities as rape, incest, homosexuality, bisexuality, group orgies, midwifery, surgery, urolagnia, suicide, matricide, murder, cannibalism and other kinds of violence and antisocial behavior.

At the other end of the spectrum are animals that employ tender passion, generosity, tolerance, beauty, faithfulness, dance, song and a host of other agreeable methods to express their sexuality. Between these extremes is the majority of animals which utilize some aspects of both. In short, each species "does its thing."

There are over a million known species of animals in the world with possibly another million, mainly insects, yet to be discovered. Basically, different species do not interbreed; on the rare occasions that individuals of closely related species do mate, these unions are always sterile. Nature has seen to it that there shall be no contamination or "mixed marriages" between species.

With such a staggering variety of animals, most of which are dissimilar in physical structure, size, habitat (on land, in the earth, in or on the water, in the air), modes of survival, intelligence, inheritance and many other distinctions that apply, it's small wonder that sexual behavior differs among all groups. Sometimes the difference is slight; with others, the difference is so sweeping that the behavior is not even recognizable to other animals.

Sexual behavior takes in far more than the copulatory act itself. It includes the myriad forms of courtship that precede copulation. It involves the sexual signals that bring appropriate mates together, at times from great distances, by employment of one or more of the senses. It deals with mating seasons and estrus periods, planned meetings and chance meetings; if such meetings are impossible, the participants will release their sperm and/or eggs in the water or on land, leaving eventual meeting and joining of the two to serendipity. It all adds up to the universal verity that all forms of life are distinctive, and therefore mating habits are dissimilar among species.

To regard the mating behavior of any other species as strange or deviate is to ignore the basic fact that there is no other way to go for those animals. Either they do it the only way they know how—or die. Even nature, which designed the behavior in the first place, cannot change it except in the slow, orderly process of evolution.

A sight not entirely unfamiliar to us, that of two humans copulating, would be something to puzzle many animals (if they cared enough to wonder) and even some humans. Many a child has unexpectedly wandered into his parents' bedroom and surprised them in the middle of a sexual engagement. Often, to the young child, his parents seemed to be in a violent struggle and a common conclusion is that "daddy is hurting mommy."

Usually, the parents' feelings of shame, guilt or anger at being discovered in the sexual act pales alongside the child's traumatic shock. That's because of the secretive, uptight attitudes toward sex held in our society, in which the child is "protected" from sexual knowledge until he or she is "ready." To many parents, that would be the night before the wedding. Fortunately, most children, in many ways wiser and more natural than their parents, don't wait that long.

This rigid, fearful attitude toward sex has narrowed the range of acceptable practices in this area and placed all other sexual variations beyond the pale. Such determinations, accepted by the community, are dictated by the "authorities" on sex. (An authority is anyone who says he is one, but usually a representative of the judiciary, police, legislative or church sector of the society.)

A common example of such sexual bigotry, until very recently, was our attitude toward masturbation, particularly among children; masturbation was for generations described as a sinful act. Children have always delighted in handling their genitals simply because it feels so good; the reason it feels so good is that nature planned it that way. Yet this practice was usually called, "an unnatural act," prior to the sexual enlightenment of recent times.

Now most authorities on sex agree that masturbation during youth acquaints the individual with his natural sexual feelings and actually contributes to a more responsible and satisfactory sex life in adulthood.

The same narrow, hard-nosed sexual censorship that once applied to masturbation is still applied to such sexual diversities as unusual sex positions, oral sex, anal sex, group sex, fetishes and any number of harmless sexual activities.

All this points up the way we've fought against accepting the unique and privileged position we hold in the sexual area. Only we, among all species, have not been directed by nature into a rigid pattern of restrictive sexual behavior as have almost all other species.

Just so long as we, in the end, manage to join sperm with egg, nature doesn't give a hoot as to the methods employed. It doesn't matter if we do it while standing, sitting, kneeling, crouching, lying (on the back, front or either side); on land, in the water or in the air.

Evolution isn't concerned with our minor aberrations and neither, for that matter, is our species. Just picture for a moment the most attractive and distinguished man and woman that you know. How can you be sure that either of them hadn't been conceived in the back seat of a car traveling sixty miles an hour, or behind a bush in a public park, or in the washroom of a cocktail bar by drunken parents—or under any other preposterous circumstances?

The pity of it all is that our society is engaged in rejecting one of nature's greatest gifts to humankind, an unrestricted capacity to enjoy one of life's most natural, delightful and fulfilling pleasures—sexuality.

Of all the animal species on earth, only the human is blessed with all of the following sexual freedoms:

- Constant access to sexual expression, not restricted to precise seasons or brief estrus periods;
- Freedom from the dictates of hormones, and the ability to match our sexual desires with our moods and feelings;
- Choice of various sexual positions, instead of being frozen into a single, unchangeable posture; for example, we alone can make love face-to-face and see the growth and expression of ecstasy in the eyes of our partner and share in kisses on the lips;
- And, we alone are able to multiply our physical pleasure with meaningful spiritual love and emotional feelings (though the physical sensations alone frequently are sufficiently gratifying

in themselves) that make the experience completely fulfilling.

In brief, rather than being locked into prescribed instinctual and unalterable modes of sexual expression in all aspects of sensuality and sexuality, we have many choices.

The key word here is "choice." The absence of choice in any endeavor is tantamount to repression and slavery. The existence of choice is a vital part of freedom, whether in the area of politics, economics or sexuality. In human sexuality, choice is boundless. The fundamental choice is one of attitude. One who chooses to repress his sex drive and instincts will always be a reluctant and frustrated prisoner of his sexuality. One who chooses to accept and respect these natural feelings and instincts will find liberation and exhilaration in his sexual life.

All this leads to the question: Why do we insist upon repudiating such a precious inheritance? Instead of graciously accepting these special sexual gifts from nature, offered only to the human species, some "authorities" have been busily engaged for the past few thousand years in an effort to modify our natural sexual birthright. This has produced countless laws, canons, regulations and traditions, all directed toward controlling our sexual practices and stifling our natural impulses and instincts.

As is often true in such a crusade, facts and truths are turned upside down. Instead of "acting like animals" when we follow our instincts, we're truly "acting like humans." Only if we succumb to demands for rigid conformity in our sexual behavior will we be "acting like animals."

This is why all the attempts to contain and mold our sexual expressions have always failed in the long run. Even the combined power of the state, church and societal institutions is no match for nature in the arena of natural instincts.

This is not to say that sexual chaos is being recommended. There's a single guideline that adequately covers the range of human sexual behavior:

> *Consenting adults, in privacy, without coercion or harm to anyone, can engage in any kind of sexual behavior they desire.*

It's not known why our species was chosen as the sole beneficiary of all these unique and protean sexual gifts, but shouldn't we accept them as freely as they've been offered, and finally stop "looking a gift horse in the mouth"?

2
IN THE BEGINNING

In the lives of some individuals the discovery of sex becomes an unforgettable moment, forever ingrained in their memories. Conversely, some find this experience of such little import that they remember the occasion with difficulty. Still others recall it with distaste or even repugnance. However, for the first group above, the climactic experience is so ecstatic, so highly charged with unexpected sensations that, for a moment, they might believe that the peak experience just savored was unique to themselves, and even that they had invented the act. That conclusion misses the mark by some three billion years.

It was way back then when the first unicell or protozoa introduced sex into this world by splitting into two. With that first act, these infinitesimal creatures learned something we were never taught in school: how to multiply by dividing.

To grasp the distance we've traveled since the first of the single-celled organisms, consider that a human being is composed of 100 *billion* cells! Yes, we've come a long way. Of course, if we each had to split our 100 billion cells every time we wished to reproduce, sex would be a dreary nuisance. That's probably the reason that nature figured out a far more fascinating and fulfilling way for us to do it. Long live nature!

Before we become too complacent, let's bear in

mind that despite our marvelous evolutionary advance we're still merely mortal. On the other hand, the original form of life, the lowly protozoa, retains the only living form of immortality as it keeps dividing itself into infinity. As a recent saying goes, "Bigger ain't necessarily better."

Also, these little protozoan rascals have it all over us when it comes to virility, fertility, fecundity or just plain sexual activity. It appears they rarely take a breather from their procreative gymnastics. There's constant sexuality as each cell splits up into hundreds of minute baby cells which become sexually mature in a few moments when they, in turn, split into hundreds of additional tiny cells. We'd be up to our clavicles in protozoa if it weren't for such obstacles as natural enemies and unfavorable environmental conditions. Even so, there are far more of these tiny beings than of all other living creatures on earth combined.

When one considers that statistic against the fact that they are so minute most of them cannot be seen by the naked eye, then the colossal magnitude of their nonstop sexual activities becomes more apparent. Come to think of it, maybe it's just as well that we can't see what they're doing.

One definite advantage they have over us in the sexual sphere is the various methods the protozoa utilize in the reproductive act. This doesn't refer to sexual positions or choice of sexual activities but to the actual methods of propagation. Different groups of these unicells reproduce by fission (division), budding (much like a plant), conjugation (resembles copulation) and, finally, the real thing, copulation. The protozoa have sexual options we've never even dreamed of.

Since the time of that original unicell there have developed on earth over one million different species of animal life. Living species and records of fossil remains compose the historical evidence of the evolutionary progression that has led to the final product—*Homo sapiens*. This process is the fundamental law called biogenesis—the development of life from preexisting life.

The human embryo, in a fascinating display of biogenesis, encapsulates the three billion years of life on earth in a mercurial, kaleidoscopic reenactment of its evolutionary history.

The cells in the human gonads upon maturity undergo cell division, at which time each cell remains with half of its original forty-six chromosomes, or twenty-three each; the cell division is a replication of the original protozoan sex act. Upon meeting and merging at conception, the male sperm and the female egg jointly contain the correct quota of forty-six chromosomes, each carrying special hereditary characteristics.

Following conception, there appears a formless clump of cells. After a brief period it forms into a wormlike shape, with further transformations occurring upon the appearances of gills, tail and digestive system before the embryo adopts its final human form.

About a billion years ago the unicells, which had developed into multicelled animals, produced the first true giant—the sponge. So perfectly developed is the sponge that it has survived the billion years in its original form. It's rare to find anything so impervious to change in nature. Given enough time, massive mountain chains wear down to foothills, seas diminish and disappear, and even continents move away from each other. Nature, therefore, must be as content with the sponge's status in life as is the sponge itself. The sponge is a true hermaphrodite, producing both male sperm and female eggs. However, it produces them at different times so there's no self-fertilization.

Instead, it releases into the water sperm cells that fertilize the egg cells of other sponge colonies; in time, the procedure is reversed. The comparison of this impersonal sex-at-a-distance technique is another indication of how far we've traveled along the páth of evolution.

Following on the lowest rung of the evolutionary ladder were the invertebrate marine animals (coelenterates), one order of which is the polyp. The polyp has a hollow cylindrical body with a single opening at one end. From them came the corals, sea anemones, and medusa or jellyfish. These latter produced the fascinating species, the Portuguese man-of-war, which gave the world the beginning of specialization.

The man-of-war is a group of polyps attached to a floating bladder with each polyp assuming a separate and distinct function: locomotion, catching of prey, eating and digesting and, of course, propagation.

Then, about 500 million years ago, evolution took a giant step forward with the appearance of the annelid, which happens to be a worm. This was the world's first animal with an alimentary tract (the eating and digestive tube) and two openings in the body, front and rear. Humans and all the higher animals have evolved from the worm.

That's not quite as bad as it appears at first blush, at least in one important area, because of the more than one million different species on earth none has a more bizarre or sensational sex life than the worm.

For openers, the common earthworm possesses both male and female genitalia; when two worms copulate they impregnate each other, thus achieving maximum mileage from time, energy and bodies. This far-out hermaphroditism pales in contrast to the sexual antics of other worms, particularly the parasitic varieties. These latter include in their *affaires de coeur* such diversions as wholesale incest, rape of newly born or even not-yet-born females, cannibalism, transformation of body parts and other eye-popping examples of sexual phantasmagoria.

Moving on to about 300 million years ago, the arthropods made their appearance on the evolutionary scene. These are animals with segmented body and limbs, primarily insects. They include the arachnids such as spiders and scorpions and also crustaceans which are lobsters, shrimps and crabs, among others.

Later arrivals were the mollusks, which are mostly shell-dwelling animals such as oysters, clams, mussels and snails; the cephalopods, or squid and octopus; and the selachians such as sharks and rays. Incidentally, the shark is not a true fish, having come along much earlier on the evolutionary scale.

Several varieties of these latter animals share a common sexual technique, the use of a gonopod, or artifical penis, for reproductive purposes. That is, the male's arm, leg or tentacle is used to replace the function of the penis.

The spider's auxiliary penis is called a maxillary palp, the squid's is known as a hectocotylus, while the shark and ray use their pelvic fins. The substitute, rather than acting as a true penis discharging sperm, acts as a transfer agent by picking up sperm packets stored in the body and inserting them into the cavity of the female, adjacent to her store of eggs. In all cases, however, before this transfer can be accomplished successfully, the male must take steps to get the female into a properly tender, loving and receptive mood.

If the promise of sex is powerful enough to turn a shark into a tender, loving creature then we can all rest assured that sex is here to stay.

The sexual behavior of spiders, crickets, praying mantises and some groups of grasshoppers is quite another world—a brutal and savage one. That's because the females of those groups are mostly *femmes fatales*—immediately upon completion of copulation they proceed to eat their lovers. When one of these ladies invites her swain to come on over for some home cooking he never asks her, "What's for dinner?" He knows.

Compared to the foregoing exciting and blood-curdling sexual practices, the sex lives of most sea mollusks are completely drab. Most varieties don't even have contact, but reproduce at a distance. The males discharge their sperm into the water and the females do likewise with their eggs. If and when the sperm and eggs meet in the sea, fertilization occurs. Despite such blind reliance upon serendipity for its propagation success, the system works. One reason is the staggering numbers of seed and eggs released so haphazardly by the mollusks. For instance, it's estimated that up to a million eggs a year can be laid by a single oyster.

It took nature over 100 million years to evolve the penis and vagina, and most of us, I'm sure, would agree that it was worth waiting for. You'll never guess in which animal group these anatomical curiosities developed! It was the reptiles. Perhaps the biblical story of Eve and the snake in the Garden of Eden had more going for it than some of us had believed. The saurians, one of a group of primitive reptiles that includes lizards, tortoises, crocodiles and the extinct dinosaurs, developed the first genitals as we now know them.

In primordial times the reptiles possessed cloacae, or anuses, which functioned as the combination urinary, excretory and sexual organ. Therefore, in order to copulate, the male and female had to unite their respective cloacae to form a tight-fitting duct through which the sperm could be transferred to the eggs. This technique required a perfect alignment of the two cloacae in order to prevent any leakage of life-giving sperm.

Such an imprecise method of fertilization, particularly among the mammoth ungainly dinosaurs, unquestionably disturbed the orderly processes of evolution, and nature set about to correct a major deficiency by modifying the sexual apparatus.

It was apparent what nature was up to when the male reptile achieved the facility to protrude his anus, and insert it into the female's anus. A further refinement took place with the addition of one or more fingerlike extensions to the male's cloaca which then had greater penetration into the female.

Finally, the possibility of the male's cloaca slipping out or disengaging itself from the female during intercourse was foreclosed when the male's member adopted a spiny covering. These spines enabled the male securely to hook his cloaca inside the female during copulation with complete assurance that the sperm would reach its goal.

The efficiency of this genital structure can be observed when a pair of mating lizards or snakes are interrupted and the frightened male hurriedly departs, dragging his erstwhile inamorata along by her cloaca.

Today's crocodiles and tortoises are survivors of the age of giant reptiles. They have utilized those millions of intervening years to modernize their sexual equipment. Both sexes of these animals possess cloacae but the male has an additional member attached to the cloaca which, when quiescent, is hidden within the aperture. During sexual arousal, however, blood flows to the genital area (precisely as with humans), swelling the member so that it emerges from its hiding place, similar in function and appearance to the erect penis; however, its channel for the sperm flow is primitive in that it is exterior instead of interior. This arrangement, the archaic penis, facilitated copulation, and it is the prototype of the penis adopted by all mammals, including humans.

It seems clear that the engineering design of the penis-vagina paradigm is functionally superior and unquestionably more pleasurable than the clumsy matching of male and female cloacae. Yet, almost all bird forms which, incidentally, evolved from the reptiles, rejected the penis-vagina model in favor of the cloaca.

There is one thing to be said in favor of the cloaca model over the penis-vagina and that is, except for rare instances among a very few species, the complete absence of rape among birds. That's understandable, considering the necessity for absolute cooperation by the female in order for the sex act to occur.

Only in a few groups of birds, such as ducks and ostriches, is rape even possible because the males of these species are the rare exceptions that possess penises. In short, rape can occur only when a penis is involved. Even then, it's extremely difficult if not impossible when the female has the ability to fly away.

3

HOW DIFFERENT ARE WE FROM OTHER ANIMALS?

As poets have written, humankind received the gift of upright stance so that in this elevated position, we would better be able to reach for the stars.

In the entire range of the mammalian world only *Homo sapiens* is immune to biological time clocks or inbred and instinctual signals in order to copulate. That's because all other animals mate for the sole purpose of procreation while humans can also utilize the sexual experience for other reasons.

It was an exquisite sense of omniscience on the part of evolutionary genius to increase the sexual options for the human species from its otherwise single and binding purpose of reproduction. Imagine how desolate existence would be if our only sex outlets were locked into propagation of the species; if our sexual expressions were communicated only during brief, prescribed seasons; if we were denied a free and unrestricted sensuality and sexuality as the vital core of our man-woman relationship!

Certainly, without this self-determination and freedom in sexual expression, the human race would have constructed a kind of society entirely different from what we have at present. If our sex outlets had been confined solely to the exigencies of sexual seasons then perhaps, like our closest relatives the anthropoid apes, every female in estrus would be fair and willing game for any and all available males.

Under such a biological regimen, the nuclear family, the bedrock of our present culture, probably would never have been invented; it would be impossible to determine the true sire of any offspring. This might have led to blood groupings, the extended family unit or clans, comprised of children, parents, grandparents, cousins, aunts, uncles, nieces and nephews, all engaged in anarchic polygamy and polyandry. Incest, in such a society, would be not only unavoidable, but rampant.

Many would argue that intelligent and moral beings such as ourselves would never permit such chaotic sexual practices, at least not for the entire human race. That may be true, but if it were possible to trace our lineage back to the time when the females of our prehominid ancestors were still captive to periodic estrus, perhaps a few million years ago, we might find that the extended family unit was the norm.

Even today, if the human female's sexuality were determined by her periods of estrus, it's highly doubtful that man, or woman, would tolerate sexual abstinence for three fourths of the time, or more. Under such conditions, extramarital activities such as adultery, open marriage, wife swapping, ménage à trois and group sex would be pandemic, rendering marriage a meaningless travesty.

Another social possibility is that, given our high intelligence, organization and desire to accumulate possessions, we might have evolved into an absolute matriarchy. In such a society, since the mother would be the only known connection with all offspring, she would become not only the biological matrix but also the source of all lineal descendancy. Thus, she would be the one to collect property and pass it on to her heirs, probably only to females.

Yet another scenario might have been a communal society where individuals owned nothing but shared in everything, including females, much like the higher primates. With unlimited human ingenuity, very likely other forms of organized society could have developed.

Nature endowed us with the ultimate sexual gift—open season for sex with (practically) no strings attached. Of all living creatures, apparently we are the only ones that require this facility in order to survive, at least with our sanity intact. The same marvelous symbiosis of intelligence and feelings, pragmatism and emotionalism, that brought human kind to its present preeminence has, along with its progress, placed certain unavoidable demands upon the human organism.

Our unique faculties of high intelligence, memory, knowledge of the past and awareness of the future including the certainty of death, an active conscience, plus the reactive feelings of guilt, anger, spirituality, fear, hate and love, are all qualities that testify to our humanism but also bring a certainty of tension to the complicated area of our interrelationships.

This is particularly true in the primary relationship of a man and woman who come together to remain in a permanent union which results in offspring to secure the continuation of the species. This bond produces feelings in both participants that are stronger and deeper than those found in any other liaison. These feelings, called "love," are constructed of various parts such as admiration, respect, trust, liking and, of course, physical desire to hold, kiss and stroke. The combined

thrust of these feelings and intimate contact lead inexorably to the ultimate physical intimacy—sexual intercourse. The unique capacity for coitus at any time is not only admirably suited to the complicated and demanding emotional and physiological structure of the human being but is an absolute necessity for our survival. Without it, the human race would long ago have succumbed to a fatal malady brought on by intolerable sexual frustration and boredom.

Yes, humans are far different from all the other animals. On the other side of the coin, however, are the many similarities that humans share with other animals. Few of us are able comfortably to acknowledge our kinship with the lower orders, yet, how can we deny this, since we all come from the same beginnings?

Going back a mere three or four million years (a very brief time in paleontology) would find the first hominids (human ancestors). At that time, our progenitors were eighty-pound, four-foot-tall creatures which had just invented the first artifact—a small, sharp-edged rock, used as a weapon and a tool.

Backtracking another 12 million years would find us represented by the anthropoid apes and at a point where our ancestors were about to branch out into our prehominid phase. Between that period and 45 million years ago, our precursors were monkeys and anthropoid apes which had evolved from the original primates.

Retracing our descent still another 25 million years to about 70 million years ago, about the time that the dinosaurs became extinct, primitive primates first appeared—tree-dwelling, shrewlike creatures. To get an approximate idea of what our original ancestors were like, look up a picture of a lemur, a direct descendant of that first primate.

Though tens of millions of years separate us

from those early days, our instinctive brain, the residual collective unconscious mind, shares many instincts and evolutionary memories with numerous other animals. This sharing of identities applies to the basic emotions and functions, including the sexual.

Many animals behave during courtship very much like humans; or is it vice versa? Many of the mating ceremonies of various animals, some of which are described elsewhere in the book, are esoteric and far beyond the experience of most of us. Despite the infinite range of distinctions among the more than one million different species, beginning with the one-celled amoeba and ending with the 100-billion-celled *Homo sapiens*, some of the sexual behavior described does ring a bell; it's familiar because of its close resemblance to the human sexual experience.

The elephant is one of the least likely of all animals to appear like a human in love, yet these six-ton behemoths behave very much like our own adolescents caught up in the throes of a love affair.

When the female elephant comes in estrus, she selects a mate, and this marks the beginning of a very close, affectionate relationship. At first, she's very coy and flirtatious, alternately inviting the bull's advances and then running away from him. As the romance progresses, playfulness turns to tenderness as they lovingly touch, bathe, and feed each other. During the courtship of several months, they're inseparable. Every waking hour is spent together, playing, touching, stroking, petting and mooning. As the affair deepens they become more intimate, using their trunks in erotic play as effectively as a pair of human lovers with four arms and twenty fingers. The male displays remarkable restraint and only at the end of the long courtship, and only at the female's invitation, does he consummate the relationship in copulation.

Some animals react in a very humanlike way to certain external influences. For example, does the full moon have an effect on your romantic psyche? Does that pale, luminescent orb hanging overhead make your heart overflow with tender ardor, send the blood racing through your veins in a flood tide of passion, cause your very skin to tingle with sensual anticipation? To many, that might sound like sheer purple hyperbole, but there's at least one species that might consider it understated. It's a species of marine worm that makes love *only under a full moon*.

Undoubtedly, there are many among us who find no special magic in a full moon and, in fact, prefer darkness for their amatory experiences. If so, you're not alone. In one variety of marine bristle worm, the female, when ready to mate, emits a steady light in the water while the male cruises around with his lights flashing and blinking. When the two meet and begin to mate, all their lights go out instantly.

Some fish, during the courtship rituals, will kiss each other on the lips just like any ordinary human couple. Some fish seem to be unusually enthusiastic about this exercise and a pair of labyrinth fish have been timed holding such a kiss for 25 minutes. That may not be a world's record but there aren't many who could equal it: holding a kiss for 25 minutes—underwater.

If you're a woman, how would you like to have Beau Brummell, Lothario, Sir Laurence Olivier, Sinatra and Astaire (each in his prime, of course) all rolled up into a single, gorgeous male who would be available to your whims whenever you were in the mood? A silly fantasy, you say? Well, if you were a female belonging to certain groups of jungle fowl such as New Guinea's bird of paradise or Australia's lyrebird, that's the sort of dreamboat that would be in store for you.

The males of these genera sport the most brilliant, splendiferous regalia in a stunning range of colors, plumage and trains. If that weren't breathtaking enough to bowl over any female, they're also accomplished actors, singers and dancers. The lyrebird is even a great imitator. But, when mating season arrives, any female of the species can waddle up to the most fabulous, supersexy male and have her way with him. As is true everywhere, sex is the great leveler.

Compared to humans, most animals are limited in their communications. Nature, never leaving anything important to chance, has devised a set of unmistakable sexual signals for each species. One of the simplest and most cunning of such signals is odor. When a female is in estrus she emits a distinctive odor, usually via her genitals, which spreads the good news to all compatible males in the vicinity. The male response is immediate and enthusiastic.

These sexual scents are infallible messages. Significantly, some of these natural fragrances such as musk, castor and civet are the bases of many perfumes in popular demand within our culture. Is it possible that these scents are worn by today's women (and men) for reasons similar to the original purpose? If you're wearing a perfume based on the essence of musk, don't be frightened if you happen to notice that you're being followed by an alligator. It just so happens that musk is the sexual aroma that turns him on. He has no intention of harming you. All he's interested in is . . . Come to think of it, maybe you should be frightened.

Many animals have sexual attitudes and approaches to which most humans can relate.

Crocodiles and frogs are direct and abrupt, even brutal, in their sexual encounters. The males of both genera waste no time in preliminaries. During mating season they fiercely attack the females and seem to force them to engage in sexual intercourse.

Some drakes and ducks, even when enjoying a permanent relationship, are notoriously unfaithful to their spouses. During mating season adultery and cuckoldry are rampant as the males spend much of their time and energy in chasing after strange ducks. Between mating seasons, however, it's the females of some species who become sexually aggressive, making demands upon their mates which the abashed males are unable to satisfy.

River crayfish and minks seem unable to conduct their sexual affairs in a peaceful and harmonious way. Instead, they always indulge in lots of apparently violent physical attacks as part of their lovemaking. The opposite of this approach is provided by the birds of paradise and bowerbirds where the males offer beauty, talent and even homes they've constructed to prospective brides. With all their efforts, these loving birds remain at the mercy of the females who are the final arbiters of all romantic liaisons.

Lions, flies and cockroaches are supersexual athletes, as the priapic males and nymphomaniacal females, despite tremendous numbers of sexual contacts, seem to be in a perpetual state of concupiscence. While the lion is quite faithful to his pride of lionesses (with their incessant sexual demands during mating season, he doesn't have much choice), flies and cockroaches are utterly nonselective in their choice of mates.

Whether other animals receive intense pleasure from the sexual act, as humans do, is something difficult to determine, because our communication with other animals is so limited. They're unable to tell us what feels good, but maybe that's not so different from many humans who are unable or unwilling to talk about their sexual experiences.

However, it's possible to draw conclusions as to whether animals find pleasure in the sex act, by simple observation. When a female is in heat, we know that the males of the species respond with great enthusiasm, currying favor with the receptive female, jockeying for position, and displaying considerable fervor in mounting her. We can safely assume that these males wouldn't be so eager and insistent if there weren't something in it for them.

In fact, these male animals perform in a manner not unlike the human male in a similar sexual situation. Scientific measurements of dogs, rabbits, rats, monkeys and other animals prior, during and after copulation, have endorsed this view; they have shown responses similar to those of humans. The internal reactions show an increase in the pulse rate, heart beat and blood pressure as the animal goes from quiescence to arousal to climax, and a corresponding recession in these measurements as the sexual excitement ebbs.

The sexual sensations of the female animal are quite another matter and much more of a mystery. In most couplings the female stands, crouches or lies passively as the male assumes the dominant, active role. However, there must be some force that turns a passive female into a willing, even insistent, sex partner. Could that force be merely blind instinct dictated by nature for the perpetuation of the species, for which the female receives no physical reward? There is some evidence to indicate that this probably is not the case. For one thing, the female animal shows the same kind of increase and recession of pulse rate, heart beat and blood pressure during a sexual contact as does her male partner.

Further, many female animals possess the primary organ of sexual arousal, the clitoris, which undoubtedly serves the same purpose as it does in the human female. In some animals the clitoris is so prominent it's frequently mistaken (by humans) for a penis. The female hyena, for example, has a long clitoris situated alongside a clump of tissue so that the combination resembles the male's penis and scrotum. In addition, the female is larger than the male so that the two genders are frequently confused. That's of little consequence, unless you happen to be another hyena.

Those species not equipped with the penis-vagina fixtures usually perform the copulatory act with much less exuberance. Female birds, for example, do not possess a clitoris and most birds mate by pressing their two cloacae together for a brief period. It's all done with little fuss or feathers.

The propagation techniques of most species of fish are even more humdrum. In most cases there's little or no contact between the two; the male merely deposits his sperm over the eggs after they've been released by the female. After that brief interlude, the lovers may go their separate ways, probably never to meet again.

However, in many species of fish a pair-bond does exist whereby a couple remains together to conceive and raise offspring. Also, in a few species that are not so aloof and distant in their mating procedures, the couple will have physical contact, even including penetration, as part of the procreative act.

While many of these sexual acts seem lackluster and unrewarding to us, there's no discounting the billions of procreative acts that take place each day on the planet. Despite the lack of emotional passion, body gyrations and even clitorises, there is agreement among many scientists that some sort of physical gratification probably does occur. As mating time approaches, tension builds up in the animal's body and reaches a peak immediately prior to copulation; this act then releases the tension.

The ensuing relaxation of tension must be a great relief and extremely pleasurable, as we all know from our own experiences. Therefore, the enjoyable sensation of relief is not localized in the genital tract but is dispersed throughout the entire body.

After mating, many animals react very much as humans do. Some lie unmoving, in a state of exhaustion; others remain very quiet and peaceful as they tenderly caress each other. It's safe to assume that nature incorporated pleasure into the sexual act of most animals, in order to ensure the perpetuation of the species.

There are a great many other examples of animal sexuality described in this book that have their counterparts in human sexuality. Many of us, indignantly or even with horror, will refuse to accept the thought of such relationship. But to deny this is to deny our basic instincts, not to mention our very source, and it closes the door to greater understanding of where we come from, who we are and why we act, at times, in such puzzling and unpredictable ways. Whether we accept or deny our sexual heritage, it won't matter; nature will prevail.

Having examined some examples of animal sexuality that clearly indicate an affinity to human sexual practices, we should also look at other examples of subhuman sexual behavior that ring a fainter bell; those that are on the periphery of human sexual behavior and go under the heading of aberrations. They are considered perversions according to our civilized mores and culture, but to the animals involved they're simply normal, instinctive, down-to-earth sex with no other options available.

We must never forget that the human species traveled along the very same evolutionary road, and at various road stops in the distant past we undoubtedly carried out procreation by some of the same sexual practices now extant among the various lower forms of life. Basic residues of these archetypical sexual experiences remain hidden in the human collective unconscious; when they surface as instinctual impulses, they invariably are branded by society as perversions.

In the same context, the self-appointed guardians of our morals are quick to designate someone as an "animal" when they disapprove of his sexual conduct. Yet, since an animal *always* follows its instincts, which are directed by nature, how much purer or more innocent could its actions be? or, by extension, the actions of the "animal" in man?

As a simple example, oral-genital sexual contact is very common behavior among almost all animals, including primates. We've all seen dogs sniffing and licking each other's genital area. This behavior is too consistent for it to be dismissed as a meaningless, dirty, canine habit. It has tremendous significance for the dogs involved. For the sniffing dog, it firmly establishes the other's gender and, if female, whether she's in heat and therefore sexually receptive. This is all part and parcel of evolution's grand scheme for the survival of the species.

At some period back in the dim, misty ages, our distant progenitors were innocently engaged in this same kind of normal, oral-genital, sexual behavior. Could our unconscious memory of our evolutionary origins be the cause of the current, growing popularity of oral-genital sex?

This practice is not only as old as the human race but, in fact, much older. Interestingly, the oral-genital technique has become so widespread in the present permissive sexual climate that it has resulted in new sexual codes enacted in our legislatures.

Another subject of even greater abhorrence to the righteous defenders of our sexual mores is anal sex, which is considered to be so heinous that it is considered a criminal offense in most countries. Yet, the millions upon millions of birds and other animals that mate—perhaps even as you are reading this—do so by firm contact of their cloacae, or anuses. And we can assume that nature has provided the participants some physical delights and satisfactions, integral to the act.

Somewhere in the depths of our unconscious may be vestigial memories of such primeval sexual pleasures. This may be a way of explaining the incidence of analinctus and anal intercourse or pederasty. Certainly, the anal area is richly endowed with nerve endings and is one of the erogenous zones in almost all humans.

It is probable that in our evolutionary progress our ancestors have experienced sexual practices later considered sexual perversions. The portrayals in this book about animal sexual behavior, whether familiar or bizarre, may help you deal with this speculation. You may be surprised at your own feelings and reactions in response to the sexual revelations of the animal world.

4

SEX—BIZARRE, FANTASTIC, BUT COMPLETELY NORMAL

Perhaps the reason nature is so unimpressed with human sexual foibles and unorthodoxies is that, compared to some other animal species, our most bizarre sex practices are tame indeed. Perhaps, as we learn about some of the incredible (to us) sexual performances of other creatures we will view our sexual abnormalities with a less jaundiced eye.

THE EARTHWORM
Homosexual? Bisexual? Yes

The earthworm is a complete hermaphrodite, possessing both male and female genitalia; when two worms copulate, they impregnate each other. That was a wise move on the part of nature because if you've ever seen a worm, even close up, you know there's no way to determine its sex. Since worms spend most of their lives underground in pitch blackness, even the worms might have problems sorting each other out. This way, when two worms meet, they can just go at it without bothering to check whether the other is a boy or a girl.

As simple and felicitous as this arrangement seems it's really quite a complicated procedure. To begin with, when two worms meet they snuggle up together, lying side-by-side. They maintain this position by the collars that girdle the centers of their bodies; the collars stick to each other. After some time of this intimate body contact, a kind of double copulation ensues as each male organ erects and ejaculates into a special sperm pocket in the other's body. Following this, the two worms separate. The entire sexual union has been a strictly homosexual affair as only the male organs have been involved.

That episode is only the opening shot, with further progress in the reproduction cycle requiring solo efforts on the part of each individual. Some time after the encounter, chemical changes in the worm cause the collar around its body to swell, at the same time filling up with nutrients.

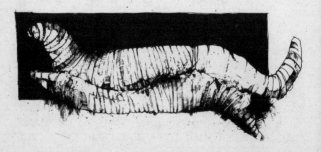

The worm, by muscular action of its body, maneuvers the collar forward until it reaches the egg cache. The worm transfers the eggs into the collar. Continuing its muscular gyrations, it moves the collar further forward to the sperm pocket and releases the sperm to join the eggs. The worm's final contractions move the collar completely off its body when it becomes a cocoon for the fertilized eggs.

After the eggs hatch they feed on the cocoon's nutrients until they're ready to crawl out as full-fledged young worms, soon to participate in their own reproductive activities.

It's another example of nature's infinite ingenuity where-by a seemingly sterile homosexual union turns into a fertile reproduction process.

THE THREADWORM
Love Is Where You Find It

Threadworms infest various grains and vegetables such as turnips and potatoes. The female is larger and has its vagina in the center of its body while the smaller male has a penis in the shape of a forklike hook. While the male and female will inhabit the same potato, they don't really get to meet each other until the end of an unusual assignation; even then, he doesn't get to see her face.

When ready for fertilization the female pokes through the outer skin of the host vegetable but only enough to protrude her vagina. At the same time, the male leaves the interior of the vegetable and explores the surface until it finds the object of its search. It then inserts its hooked fork member into the exposed vagina to complete its sexual odyssey. This is something to ponder when one is scrubbing a potato.

THE AFRICAN BLOOD FLUKE
Sexual Menace

Another parasitic worm, much more harmful to humans, is the African blood fluke, though its sex life is extraordinary. The skinny female starts things off by crawling inside the male's larger body where she finds not only lodging but copulation that's constant and ongoing. That is, until she's ready to lay her eggs, at which time she leaves her cozy nuptial chamber and travels to the pelvic area of the human host. Here she deposits her fertilized eggs which are flushed out in the human wastes to end up, frequently, in some river or other body of water.

The eggs may then be ingested by unsuspecting humans in whom the fluke's reproductive process will be repeated. These parasitic trysts create havoc among thousands of Africans, causing general debility and even death—a heavy price to pay for the singular sexual peccadilloes of the blood fluke.

THE BUMBLEBEE EELWORM
The All-Encompassing Vagina

The bumblebee eelworm, which receives its name from its habitat, the body of a queen bee which it infests, leads a sexual life that makes any horror chamber look like a nursery.

For openers, it copulates in the wet ground, following which the male dies (and probably not from pneumonia). Now impregnated, the female seeks out a queen bee host in which she makes herself comfortable for the accouchement.

At this point, a weird development starts to take place. Her vagina begins to expand and continues to grow until it has absorbed the entire genital tract—the womb and ovaries. This abnormal activity does not stop there; the vagina continues to grow and encroach upon the remainder of its being until the body shrivels up and drops away. The worm now is one hundred percent vagina which leads its own life until the young worms have hatched out of the eggs.

THE CARP PARASITE WORM
Permanent Love Connection

The love life of these parasitic carp worms is quite unusual in several aspects. To begin with, they spend their entire life together in underwater travel. That's because they set up their permanent residence on the gills of certain species of the carp family. Even more unusual is their arrangement of sex organs as each individual is endowed with two sets of genitals, one male and one female. Now comes the most unusual feature of all: when two of these creatures come together to mate, once joined, concrescence sets in and they're stuck together for life.

This union has to proceed rather carefully since there are twice as many parts that mesh than is usually the case. The male genitals of one must be attached to the other's vagina while at the same time, its vagina must correspond to the male parts of the other. So, for their entire life together, all four sexual organs are permanently connected.

THE STARWORM
It's Good Work, If You Can Get It

The female starworm is in no danger of winning any beauty contests. She's shaped like a pickle, has a sickly green complexion, and is covered with warts. Despite her ghastly appearance, she has no lack of lovers. In fact, there are always a half dozen or more in full-time attendance.

One reason that her unprepossessing looks don't turn off her male friends is that they never get to see her; and vice versa. That's because they're diminutive dwarfs that set up permanent residence inside her vagina. They're sort of live-in lovers whose primary function is to fertilize the female's eggs as they're released.

This cozy life-style for the males is completely self-contained; they don't even have to send out for food as they take their nourishment from the nutrient juices of their hostess.

THE MARINE BRISTLE WORM
Castrating Females,
or Watch Out for That Love Bite

Animals, even on the very lowest scale of evolution, may display primitive examples of courtship behavior. One of the simplest and most elementary of these takes place among certain marine bristle worms which, during breeding periods, congregate in large numbers. The males perform a lively dance during which they undergo extreme contortions. This *corps de ballet* succeeds in driving its female audience into a sexual frenzy which, at a climactic point, will induce the females to drop their eggs. The dancing will then stop as the males cover the eggs with their sperm.

In the platynereis genus the female will get so turned on by her partner's erotic dance that, finally unable to restrain herself, she'll lose all control. Consumed with sexual excitement, she'll bite off the sex organs of the male which are located in his rear end. (A kind of endless sex.) She swallows the sperm contained in this mouthful, and the sperm travels through her body to the cache of eggs and fertilizes them.

THE DEEP-SEA ANGLER
Her Lover Grows on Her

The female deep-sea angler is quite large. The angler inhabits the ocean at such great depths that she uses a light to attract prey and lovers. The male is a dwarf, about one twentieth the size of the female. He also is quite agile, far more so than the cumbersome female who otherwise might consider the tiny male more tempting as a meal than as a lover. The male cruises around until he finds a female ready to spawn. At this, he uses his pincerlike jaws to attach himself to the appropriate part of the female and fertilizes her. In some cases, after fertilization the male leaves in search of another pregnant female.

In other species of this fish, however, several males band together and use their pincers to attach themselves permanently to the female. In time they grow partway into her body. In this position the males feed directly from the female's bloodstream, and they are always on call for sexual duties.

THE GUINEA PIG
Inventor of the Chastity Belt

The domesticated guinea pig is a rodent, similar in appearance to a small rabbit. It is quite sluggish in its movements. Perhaps the fact that these animals are used primarily as pets or for laboratory experiments, with the attendant care and security, contributes to their lethargy. We can assume that their cousins living in the wilds of South America are far more active in order to survive.

This difference in rate of activity does not apply when it comes to sex. The wild guinea pig mates only once a year and normally produces two offspring. The domesticated ones breed about three times as often and deliver up to a dozen times as many descendants. They seem to reserve more energy for sex and less for other activities.

The female usually comes in estrus in the middle of the night (this might explain some of their lethargy—they don't get enough sleep). The male ascertains the female's sexual condition by first sniffing her mouth and then her genitals. If he gets a receptive signal, he mounts her and that's that.

However, the female frequently may not be sufficiently aroused and will move away when the male attempts to mount her. He'll follow her. Every time she stops, he'll try to mount her, but she'll move off again. This stop-and-go series can continue for some time, until she's ready for copulation and assumes the sexual crouch; or the male may finally lose interest and wander off in search of easier pickings.

Frequently, even when she crouches invitingly and he mounts her, she may move away at the last second. This maneuver may be repeated a number of times; whether she's a tease or simply can't make up her mind is something that only she knows. If these near-misses succeed in frustrating the male sufficiently, he may rise on his hind legs, penis erect, and urinate over the female. Occasionally, she may find this action stimulating and submit to his sexual demands. More often, this will annoy her and she'll proceed to clean and groom herself.

When mating does occur, the male ejaculates a large amount of semen which quickly coagulates into a hard plug, effectively blocking the vaginal passage for some time. Its primary purpose is to ensure impregnation by keeping the sperm from being lost. Another effect is to prevent the female having further copulation for, by the time the sperm mass has dissolved, the pregnant female has no further interest in sex—at least for a while.

Despite his sluggish ways, the guinea pig seems to have succeeded in inventing something that some male chauvinists are still searching for—the perfect natural chastity belt.

THE BEDBUG
No Vagina? Drill Your Own!

One of the most repulsive of insects (to humans) is the bedbug, with its foul odor and penchant for feeding on human blood. It's true that this bug will attach itself to any warm-blooded animal but only if its favorite feeding trough, a human body, is not available.

When it does go on a feeding binge, it doesn't stop sucking until it's completely sated and bloated. It will then refrain from feeding until it's completely empty again. There are no between-meal snacks for this insect.

There is one positive thing that can be said about the bedbug; it does not transmit disease. Despite its offensive odor and repulsive habits, it's clean.

Once having partaken of blood, the female's swollen body begins to manufacture eggs and is ripe for sex. In the mating procedure, the male's penis penetrates the female's abdomen. There's an obstacle here that would present an insuperable problem to a human. The female *possesses no sexual opening!* The male responds to this by making his own opening! He doesn't go about this in a hit-or-miss fashion, carving out a vagina wherever it may suit him. Such thoughtless abandon could easily result in the female being stabbed to death. Instead, the male finds a notched outline on the female's abdomen which guides the male member to the desired spot for penetration. The male's penis is admirably suited to such forcible entry, being large, curved and sharply pointed.

Having made penetration, the male discharges an abnormally large quantity of semen into the female. The semen swim around in the female's blood until they arrive at special sacs, from which they swim up the egg tubes to the ovaries. Here they come to rest and wait for the eggs to come along to be fertilized.

After intercourse, the female's sexual wound quickly heals, leaving a scar. (Presumably, one could turn over a female bedbug and by counting the scars on her abdomen, determine the extent of her sexual experience or even if she were still a virgin.)

The large quantity of semen injected by the male might seem wasteful but the female uses some of it as nutrients. This unusual method of "intravenous feeding" would certainly be useful during periods when blood was scarce. In fact, there's one species of bedbug which finds the males piercing each other and copulating. Rather than indulging in homosexuality, the males are feeding each other with their semen, which is rich in nutrients.

THE BEDTICK
Don't Stop Feeding
Just Because We're Copulating

While not related, the bedtick shares some important similarities with the bedbug. They're both tiny insects that feed upon the warm blood of mammals. The social manners of the tick are much grosser than those of the bedbug which, at least, has the good grace to separate its functions of feeding and copulating. The tick observes no such niceties and invariably combines the two, making it all a bloody affair.

The male approaches the female after she's attached herself to a warm-blooded body and forcibly wedges himself between his mate and her host. He then pushes his proboscis into her vagina which he proceeds to dilate. Following this, he uses his proboscis and feelers to insert a packet of sperm into her expanded vagina; during all this she never stops feeding. It all has a certain cold-blooded (warm-blooded?) efficiency as the intake of the nutrient-rich blood aids in the development and fertilization of the eggs.

5

GROUP SEX

SOCIAL IS SEXUAL

There are some animal groups such as moles, crickets, grasshoppers, leopards, bears, tigers and many others that are loners and spend most of their lives in isolation. But, when mating season rolls around, they instinctively find each other for the progenitive act.

However, most animals are social beings and are part of a group because such membership is beneficial to the individual and, therefore, to the species. For example, fish feed better when in schools and cockroaches are more secure and find their way around more easily when in small groups.

The original and basic purpose of social life is sex. For one thing, mates are more readily available in such a milieu. Of even greater importance is the effect of mood transference as sexual feelings are transmitted by various individuals of the group to all the other members in a constantly rising intensity. Upon reaching the necessary pitch, the entire group will be swept up in a wave of sexual arousal, running the gamut from discreet pairing off to group pandemonium.

The great brooding colonies contain thousands of birds that congregate at breeding grounds. Their preparations for mating may begin weeks in advance as they exhibit predetermined courtship displays. These group rituals stimulate the individual birds and produce a gradual increase in the hormones of their pituitary glands.

This procedure is essential for the birds' success in fertile mating. Large colonies are far more conducive to the desired heightened excitement as the individuals react to the rising sexual tension of the group. In small flocks, where the building of stimulation is limited, many of the birds may fail to achieve the state of fertility.

In a group setting there is no specific attraction between individuals. The magnet is the group and the individuals are anonymous. There is no organization or leadership. The larger the group the more homogenized it is and the less individualistic is each member. This is the most primitive form of social community; just as there is no attraction between members, neither is there aggression. Only when love and friendship exist among individuals can aggression also be found. A member of such a group is merely an impersonal unit that acts in concert with all other members.

Normally, if one member becomes frightened and flees, the entire group will follow without hesitation—a phenomenon known as allelomimetic action. Likewise, if one individual moves away and begins feeding, the others will automatically follow suit. The most individualistic species on earth, the human, is not immune to this form of unthinking group behavior and has evidenced such conduct in countless incidents of rampaging, mindless mobs throughout history.

When a human behaves in such a thoughtless, almost always destructive manner, he's reacting to the most primitive part of the brain, the reptilian or primordial brainstem, and bypassing the neocortex, the part that provides thought and logic.

It's no surprise, then, that group activity plays an important role in sexual behavior. When one couple in a group initiates a courtship ceremony, the entire membership will do likewise. In short order, all of the individuals in the community will be engaged in copulation, usually with the nearest neighbor of suitable gender. Affection, love, or even recognition of the mate will play no part in the situation.

However, this type of group mating does not apply to birds since most fowl practice the pair-bond in their procreative activities. If birds are members of a colony, they will benefit from the sexual excitement engendered by mass courtship, but when ready for mating, they pair off to breed and defend a small area around the nest. Most of these birds confine their relationships to the reproductive process. Usually, when the propagation cycle has been completed, they leave and go their separate ways. If they should happen to meet in the future, they probably will not even recognize each other.

THE CALLICEBUS MONKEY
Love Thy Neighbor—and Then Some

The tiny callicebus monkeys of Columbia are unusual in that they observe lifelong monogamy. Upon mating, the couple marks off a piece of territory which, from that time on, they guard jealously and ferociously. A great deal of their time is spent in watching over their property boundaries and keeping them free of trespassers.

Should a neighbor be careless enough even to step over the property line, he will be unceremoniously hurled back to the accompaniment of shrill and angry verbal abuse. Thus, every day is occupied with fights, near fights and bitter quarrels among the neighbors so that it has become the foundation of their social life.

In such a rigid, tightly guarded, monogamous culture, obviously there will be few opportunities for extramarital hanky-panky. All year long, territories and spouses are jealously watched over. Then a day will dawn that will be different from all other days. It's the time when all the female glands in the community will proclaim that the time of estrus has arrived. At this, old quarrels will be forgotten, boundaries ignored, monogamy discarded, and lifelong faithful mates abandoned.

The entire adult population of the community will throw itself into a frenzied, hyperactive, nonstop, free-for-all group sex orgy for the duration of the estrus period. The bitter animosities of the past twelve months have built up tremendous tensions among all of the individuals which will now be released in sexual anarchy as erstwhile enemies make love, not war.

One would hope that there's a lesson to be learned here—that love conquers all. But it's not so. In a few days, when the "heat" has passed, all couples will return to their own territories and to the traditional roles of abusive, disputatious neighbors for another year.

THE WATER SNAIL
Hermaphroditic Weirdo

We're all familiar with the common garden variety of land snail which, incidentally, boasts of a spectacular sex life and is described elsewhere in the book. In all, there are about 30,000 known species of these gastropods, ranging in size from giants two feet in length to tiny microscopic varieties. Snails of one kind or another are found all over the world, from ocean depths to surface dwellers, from freshwater bodies to all land areas.

While they're basically sedentary animals with their only defenses being a penchant for hiding in crevices and under vegetable matter, and a fairly sturdy shell (though some varieties don't even have shells), they manage to flourish in their various habitats. That's because they're tenacious of life and can survive without food for many months. In addition, they're prodigious breeders.

With so many species extant, it's no surprise that there are great differences among them in appearance and habits. For example, among the toothed varieties, there are species in which the members are content with a total of sixteen teeth while another species can boast of 750 thousand teeth per individual (the latter are probably identified by their dazzling smiles).

With all their many differences, there's one thing that most species share in common: they're hermaphroditic, with each individual possessing both male and female sex organs. With this kind of natural (supernatural?) headstart, the possibilities of sexual variations, innovations and permutations are so incredible that the realities far outdistance our wildest fantasies.

To cite one example, some of these androgynous *rarae aves* have the capability of reproducing autogenetically; that is, the ability of one individual to introduce its penis into its own vagina and thus impregnate itself! (This kind of eyepopping, do-it-yourself ability must certainly alleviate any boredom in their lives.)

Another species is composed entirely of

females, with the eggs developing without benefit of fertilization, a form of parthenogenesis. There are other varieties in which sibling rivalry reaches its zenith as the firstborn of the embryos eat the eggs containing their brothers and sisters. Within some groups, this savage appetite takes place even within the pregnant uterus. (Anyone for intrauterine cannibalism?)

However, even with such unbridled destruction, the damage to the generation remains within bounds because snails usually produce very large families; a snail lays up to 600,000 eggs at a time.

THE MUD SNAIL
Group Orgy in a Chain Gang

If you can picture a half-dozen snails lined up in a row, all engaged in simultaneous copulation, you have an image of the European mud snail doing what comes naturally. In the breeding season, these snails congregate in small groups and form copulatory chains.

In each chain, the first snail acts solely as a female. The second snail in line performs as a male to the first female but also acts as a female when joined by the third snail. Therefore, the second through the fifth snails operate in the dual roles of male and female, while the sixth snail, or last in line, functions only as a male. Thus lined up and locked together, the half-dozen collaborators mate as a single unit.

This no doubt produces an assembly-line type of efficiency though one can't help but wonder whether the snails at either end of the line, copulating at only half of their potential capabilities, have a sneaky feeling of being short-changed.

THE SLIPPER SNAIL
Togetherness Forever—and Ever

The American slipper nail, *Crepidula fornicata*, (the scientific name gives some indication of what they're noted for) undergoes drastic and far-reaching social and sexual changes during its life.

When young, the snail is motile, with complete freedom of movement, and it's one hundred percent male. When it reaches sexual maturity it loses its motility as it becomes sessile and attaches itself permanently to some base; at the same time, it completely reverses its sex status and becomes totally female.

Shortly after these changes, it's mounted by a male and copulation follows. There's no need for urgency—their sexual embrace is forever. They will remain locked in the copulatory union for the duration of their lives. This second snail soon is

mounted by a third snail and begins the process of turning into a complete female. This procedure of being mounted and changing sex continues until there are up to fourteen individuals in the perpendicular chain.

At this point the lower snails are the largest and female, while those in the middle segment are medium-size and engaged in changing from male to female; only the upper and smaller members are still male.

This is another example of nature's ingenuity as she utilizes the maleness (semen to fertilize the female's eggs) of fourteen young snails and then transforms them into fertile females. Still searching for greater efficiency, she joins them together to form a breeding factory that will turn out millions of fertile eggs soon to become another generation of snails.

THE PARAMECIUM
A Single Cell with
Two Ways to Procreate

Mass mood transference exhibits itself among most animal species from the human on down, even to the single-celled protozoa, the paramecium. This unicelled animal, which ordinarily reproduces through division, is able to propagate through conjugation, a sexual process that resembles copulation.

Two of these microorganisms will come together and join at the open mouths. This will induce some sexual changes in the individuals. In each, a male nucleus in a cell will travel through the mouths to join a female nucleus in the other and so combine their hereditary material.

After such conjugation, paramecia can reproduce asexually by fission for many generations. When unfavorable conditions in the environment occur, paramecia will resort to conjugation. Such threat to their survival will produce the familiar tension that triggers the act of conjugation, a spontaneous reaction that sweeps through millions of these protozoa in a flash.

Probably the earliest formation of a social and sexual bond was started by protozoa with a rudimentary society of volvox. Twenty thousand or more of these microscopic individuals will gather together in the form of a tiny green ball. (In such a tightly knit group, it would probably take only one concupiscent individual to "start the ball rolling.")

THE MAYFLY
A Lifetime of Hedonism

The Mayfly devotes its entire lifetime to dancing and copulating. This idyllic life has one major drawback—the Mayfly's life-span is less than a day. (Before its metamorphosis into a beautiful flying creature, it spends several years as a water grub.) While it spends most of the day in dancing, it reserves the last hour or two for dancing and copulating.

Both of these joyous activities are done, cheek by jowl, in a compact area with a million other revelers all doing the same things at the same time. Obviously, the Mayfly does not treat its love life as a private affair. Rather, it has a need for the densely crowded ambience which provides the security and excitement critical to its successful fertilization and perpetuation of the species.

On its one day in the sun, the Mayfly does a distinctive dance from which it never varies; it rises about eighteen inches and then floats down the precise eighteen inches and repeats this flying Yo-Yo maneuver over and over again, until the sun goes down. Late in the day, each male is joined by a female and they lock in a sexual embrace without missing a beat in the dance. Joined together, they continue the graceful ballet in tandem, copulating as they dance.

At sundown, the sexually joined pairs separate and the females fly low over the water, expelling eggs from twin tubes in their rear ends. In a couple of years, the grubs which will have hatched from the eggs will be transformed into Mayflies and repeat the same joyous performance.

THE MOSQUITO
The Only Way to Fly

The Mayfly's public and crowded display of mass mating has its counterpart with other flying insects such as mosquitoes, gnats and midges. During the breeding season, the males will congregate in an enormous swarm over some high building, tree or hill, but only at dawn or dusk as crepuscular creatures do. Many of us have seen mosquitoes perform their nuptial rites while hovering over a church steeple, certainly an appropriate site for such a ceremony.

From a distance the silhouette of a million mosquitoes, tightly massed and moving as a single unit, resembles a dark cloud, except for one thing—clouds don't hum. The love hymnal of a vast horde of mosquitoes, all on perfect pitch and in perfect harmony, is enough to start the juices flowing in a million female mosquitos resting in the nearby countryside.

The females cannot resist the love chorale. Huge numbers of them soon rise from the grass and head toward the massed bank of males. As the females near the humming multitude, males will peel off from the dense throng and immediately make sexual connections. The rosy sky will become a backdrop for an epic orgy on the wing as a million couples unite in flight.

The joined participants will fly off in tandem to some resting place where the matings will be consummated. Shortly thereafter, the makings of future mosquito orgies will be assured as millions of eggs are deposited in various bodies of water in the area.

Not all mosquito species stage airborne propagation festivals. Some are far less flamboyant although not entirely discreet. The mosquito sound with which we're all familiar resembles a "bloodthirsty buzz." This is a term that accurately describes the sound, that of the female hunting for blood. The males produce several different sounds, one of which has an uncanny similarity to the human "wolf whistle" emitted by young men on street corners to attract the attention of pretty girls walking by. In a remarkable coincidence, the very same kind of whistle is used by small groups of male mosquitoes for precisely the same purpose—to attract the attention of a pretty young female flying by. Frequently, when not receptive, the female will act very proper by ignoring the whistles and hurrying on. If the female is in the mood, she'll respond with a coolike sound and one of the males will leave the gang to join her, which he does literally and physically in midair.

Thus locked in a sexual embrace, with the female in the top or mounted position, they'll fly to a nearby bush and finish copulating. She'll then go off to lay her eggs and he'll fly back to the fellows, presumably to do a little boasting, while they all listen with mouths puckered, ready to whistle at the next passing female.

THE BAT
Doing It Upside Down

The bat is a flying mammal that is almost completely dependent on flying for its locomotion since its feet are almost useless for walking. As a group, they're extremely social as evidenced by their living together in dense clusters. They not only hang around together but also hang upside down most of the time, usually in caves. It's in this topsy-turvy position that they do their mating.

While the female hangs on to some support, the male grips the female, hanging slightly below her with his belly against her lower spine. In this position, sexual intercourse is difficult, if not impossible, unless the male has a fairly long penis that is bent at the proper angle. Of course, this is precisely the kind of penis that nature has provided him.

THE FIREFLY
Orgies in Christmas Trees

Apparently fireflies don't require great sexual stimulation for mating and are seemingly content with the dainty flickers of light which provide most of their courtship rituals, as described elsewhere in the book. However, there are species of these beetles which do put on spectacular courtship displays.

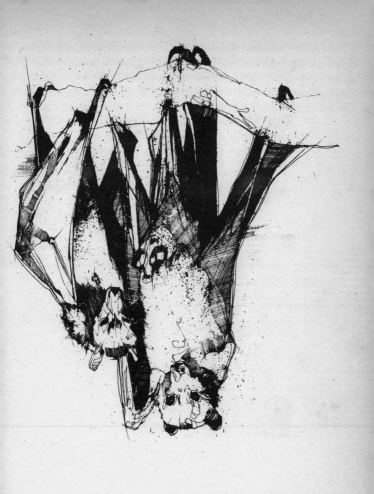

In Southeast Asia and the East Indies, fireflies congregate by the tens of thousands on the branches of one or more trees and put on a lighting extravaganza. In perfect unison, as though the many thousands of insects were all plugged into the same electrical outlet, these massed beetles flash on and off all night long, sometimes for several nights in a row. The display resembles huge Christmas trees decorated with thousands of tiny, blinking lights. The luminescence is so bright that it's visible for at least a mile away.

The light festival is completed when the lit-up males are joined by hordes of females from the surrounding jungle which are attracted and excited by the mass mating signal. Presumably, the lights aren't turned off until all of the celebrants have been sexually satisfied.

6

RAPE

IT TAKES TWO TO TANGLE

Rape, a common phenomenon among the human species, is practically unknown to all the billions of other creatures in the Animal Kingdom. Though rare, there is some incidence of this sexual aberration occurring among several of the higher subhuman primates, notably the baboon. A full-grown male may make a sexual demand upon a young female which frightens her into resisting his advances. When the male is insistent and forces the female to submit, she may become badly mauled in the encounter. However, as noted, these incidents are uncommon and are rarer yet among the other branches of the anthropoid apes.

There are many animal matings which would appear, to human eyes, to be pure and simple rape. Such rapelike actions invariably are the normal courtship rituals employed by the particular species, since mating is possible only when the female is in estrus. This condition occurs in response to hormonal changes within her body. At such times the female is not only receptive but insistent that her sexual needs be fulfilled.

With most species, the female has an inner clock that determines the time of estrus. For instance, most birds mate in the spring. For most of the year their sexual organs, or cloacae, are in a shrunken state, unfit for sexual union. In the spring the cloacae dilate in preparation for the copulation that soon will be taking place.

Among many animals the female is in estrus only once a year but some on the higher rungs of evolution have more opportunities to mate. The higher primates, for example, find the females' estrus periods approximating a monthly cycle. Among the various mammalian species on earth, only the human female is free from the dictates of estrus and can perform sexually whenever she pleases. Perhaps most human males and females aren't appreciative of this extravagant gift bestowed upon us by nature.

When estrus does make its appearance, the female animal knows it immediately and swiftly transmits the welcome news to available and eligible males. There's no end to nature's ingenuity in getting the word around to the proper parties. As marriage broker for a million different species of fauna, with perhaps another million of flora, nature has an infinite variety of signals and methods in her matchmaker arsenal; these are to assure the proper meeting and mating of all nuptial partners at the assigned times and places.

These procedures and rituals range from the flashing of lights to basso profundo choruses; from the release of fragrant aromas to sadomasochistic love-play; from wild, primitive dances to sweet serenades; from soft, sensuous stroking to cannibalism; from magical, instantaneous alteration of colors and appearances to flying acrobatics; from hypnotic trances to complete sex changes; from vicious brawls over the female to vicious brawls with the female; from incest to beauty contests; from copulations that are completed in seconds to those that go on for days; from discreet private couplings to mass orgies;

from billing-and-cooing to murder; from homosexual and hermaphroditic unions to troilism and group sex.

If there's no way to bring individuals of a species together, then their joining of sperm and egg becomes a haphazard, random effort. Even this accomplishes its objective—the continuation of the species.

Among most animals, copulation is not a specific act that occurs in isolation. Rather, it's the culmination of a series of prior actions, usually referred to as courtship, performed by the mating couple. These performances are clearly designed to increase sexual excitation on an ever-rising scale.

The building of such tension is necessary for a fertile union; it stimulates reproductive functions such as sperm emission and release of ripe eggs from the ovaries. In the absence of this prior stimulation most higher animals would be unable to copulate.

Only humans are partially exempt from this requirement. Even then, while the female is able to conceive in a passive state, the male does require sexual tension for erection, orgasm and release of sperm. From this, it appears that the human female, alone among all mammals, has achieved the ability to participate in a fertile mating without need of estrus or sexual stimulation. This makes her unique and seems to place her in the first rank of the evolutionary advance in reproduction.

Precopulatory excitement, usually aroused by courtship rituals and displays, is exhibited in myriad forms by all of the diverse animal species. To some it may appear that animal courtship equates with human sexual foreplay, and this assumption has some truth in it. Just as in foreplay, courtship has the function of stimulating the participants to the point of sexual receptiveness, if not downright eagerness. The great distinction between foreplay and courtship is that *Homo sapiens* views the former as acceptable only after some interval spent in courtship. With humans the two are rarely synonymous, whereas with other animals the two activities merge and are simply referred to as courtship.

Of course, in today's freer sexual climate, many humans are narrowing the timespan between courtship and foreplay and may have the two converge in the not too distant future.

With the exception of man, most of the higher animals treat courtship and mating as a unified procedure with the preliminaries flowing into the climax. Some species, especially among birds, conduct courtship ceremonies for some time before mating. For the most part, these are preparatory procedures or dry runs in anticipation of the real thing.

When E-Day (E for estrus) arrives, courtship and mating follow in short order. In many animal groups, the entire performance is encompassed within a few minutes; with some, even a few seconds is sufficient. However, there are some animals which require a lengthy period of courtship to reach the necessary high pitch for successful procreation.

THE MINK
Rapist or Lover?

An example of extended courtship (or foreplay) is that of the mink which has long held a popular reputation for possessing an unusually high libido. This conception may rest more on the savagery and duration of their courtship and mating than on the frequency of copulation.

From the outset, the larger male is constantly on the attack as he repeatedly attempts to mount the female while she, just as ferociously, repels him. To the unknowing human bystander, this courtship appears to be nothing less than a full-blown, long-playing rape attack. However, the minks have something else in mind. They are attempting to achieve the necessary high state of sexual arousal before copulation. In fact, the male would probably refuse to mate if the female submitted too readily. Without extreme excitation, her ovaries would probably fail to release the ripe eggs, which would thus result in a sterile union.

This whirling, snapping exhibition of furry hyperkinesis and sexual intercourse can continue for as long as eight hours. From this, it should be obvious that a full-length mink coat is something more than just the furrier's art—it also represents several hundred hours of fierce combat and copulation, courtesy of manifold and sundry minks.

THE CROCODILE
Call the Rape Squad!

An animal which has survived substantially unchanged from the primordial age is the crocodile. The male of the species is fierce and brutal in his lovemaking and, to put it simply, is the complete sexual male chauvinist.

During mating season, when the reproductive urge is upon him, the male loses no time in seeking out the nearest female and attacking her with wide-open jaws and frightening roars. Despite his wide-open jaws, it's not oral sex that he's after but, instead, he overpowers the female, dumps her unceremoniously over on her back and, crawling on top of her, pins her down in the "missionary" position. The female, thus rendered helpless, is forced to submit to his rampant sexual ardor. The ordeal is of short duration as he promptly plunges his cone-shaped organ into her cloaca and deposits his sperm.

The entire operation smacks of out-and-out rape upon a helpless victim who can't even call the police. Before any group of concerned animal lovers get its hackles raised and starts a drive to bring equality to the love life of the crocodile, there are several things to consider.

Crocodiles, as a genus, have survived fairly intact for 100 million years or more, presumably utilizing the same kind of brutal mating procedure. Recognizing that nature will not long permit inefficiency or inappropriate methods to jeopardize the future of a species, and also appreciating that even a lady crocodile is no cream puff, we can safely assume that the female crocodile tacitly approves of the situation and the submissive role that she plays. If she weren't ready and willing and, indeed, cooperative, the male would stand little chance of turning her over and invading her privacy and her eggs would not be available for fertilization. The apparent high level of violence incorporated into the mating act provides the necessary stimulation for a fertile union.

A question, idle though germane, arises. After mating, how does the male behave toward his upended sexual partner? With his lustful feelings sated, will he now show some consideration toward her and at least help her right herself? The answer is that he'll remain consistent in his boorish behavior and leave her lying there capsized, to struggle back to her feet by herself. Of course, by now his reputation is so tarnished that it probably doesn't matter.

While not intending to defend the crocodile's mating manners, one wonders if we, as a species, had as our highest purpose on earth to provide hides for fashionable shoes and luggage, how concerned would we be about life's little niceties? Even more pertinent, one wonders how many human acts of copulation, taking place daily, are not far removed from the sexual technique of the crocodile?

DUCKS AND DRAKES
Rampant Cuckoldry

In the human experience, romance, almost by definition, conjures up a picture of one individual chasing another, usually a male pursuing a female. Invariably one party is more desirous of establishing or continuing a relationship than is the other. One result of this romantic pursuit is the heightening of tensions—social, sensual and sexual—for both the pursuer and the pursued.

With humans the emotional and intellectual aspects of romantic pursuit take precedence over the physical. Among animals the opposite is true as it's mostly a physical affair, with the female coyly running from the randy male, but rarely too fast or too far. Most animals that employ the chase during courtship keep the proceedings fairly brief, with the capture inevitably leading to instant copulation.

During the spring mating season many species of male ducks become inordinately lustful (these are among the few groups of birds in which the male possesses a penis) and chase after every duck in sight. As a result, the females of these species are particularly alert during this season and spend much time and energy fleeing from the many concupiscent drakes that abound. Any duck knows better than to rely upon her mate (a permanent one, at that) to protect her honor; her mate is far too preoccupied with chasing after other ducks.

Since the duck is more agile and quicker, she can easily evade the drake unless her feelings are compatible with those of her pursuer. In such a case she'll slow down, come to a halt and, as he approaches, assume the sexual crouch.

This position is the typical sexual posture of the receptive female animal, adopted by most birds and many mammals. The female crouches with her rear end slightly elevated, tail to one side and cloaca (or vagina) exposed and readily accessible to the male.

The sheldrake displays as much lubricity during the mating season as his brethren but his philandering path is a bit more tortuous. That's because he ordinarily commences his coupling underwater when a willing female swims to the bottom of the pond or lake, an open invitation for him to follow and mount. Thus joined, they rise to the surface in tandem where the sexual act is completed. So the sheldrake's first crucial step on the way to a conquest is to persuade the lady to take that important plunge.

The mallard drake is a perfect gentleman to his mate and never forces her to submit to his sexual desires. In fact, he's quite uxorious and always waits for his partner to take the sexual initiative. During the off-mating season, in the fall and winter, her libido is usually higher than his and he's often embarrassed when he's unable to respond to her sexual overtures.

However, when spring rolls around, the mallard's juices overflow and he's his old randy self, making up for his off-season deficiencies. Not necessarily with his mate, though, since he spends much of his time, like others of his genre, frantically chasing after someone else's mate. His conscience and guilt feelings may be somewhat assuaged in the knowledge that there are bound to be other drakes in lascivious pursuit of his own mate. In the end, it's possible that this continuous merry-go-round of blatant infidelity and cuckoldry has a way of evening up things for everyone.

THE FROG
Music to Rape to

Several hundred million years ago, when the original amphibian ancestor of the frog first climbed out of the sea onto the land, it entered a world that was mute. Except, perhaps, for the cricket which had arrived on the scene a bit earlier, there were no voluntary sounds emitted by live organisms, and certainly there was a complete absence of music.

Those first amphibians soon rectified that unhappy situation and the world came alive with a cacophony of bellows, squawks, croaks, growls, bawls, yodels, caterwauls, roars, trumpets, grunts, hoots, honks and ululations. Certain groups of these frogs have long been held in reverence by gourmets for the delicate succulence of their hind legs. It is only fitting that we now pay homage to this homely creature for its enormous contribution of having brought the joy of vocal music to a desolate world.

The amphibians are a class of cold-blooded vertebrates, falling between fishes and reptiles and comprising the groups of frogs, toads, newts, salamanders and allied genera. While these are the animals that led the way out of the oceans to dry land, they still haven't made up their minds where permanently to hang their hats. Most of them live on land but they don't roam very far from the water.

In fact, most of them pass through an aquatic larval stage when we know them as tadpoles (pollywogs), at which time they have tails and breathe through gills. During the metamorphosis into adulthood they lose their tails and gills and begin to breathe through lungs and through the moist glandular skin.

Singing, to a frog, is not considered entertainment or even a cultural or spiritual event. It's a deadly serious romantic business, his only way to attract a mate. Serenading is pretty much his life's work.

If a female does not respond or if available females are grabbed off by other quicker or stronger males, he will continue his concert for hours, as anyone who has lived near a swamp or river will attest. Of course, his virtuosity and persistence will prevail in the end and he'll get his girl. Even if no one else believes this, the frog does.

The serenade is the principal part of his courtship and has the effect of stimulating both the listening female and himself. Therefore, the longer he continues his musical invitations the randier he'll become. If he's completely shut off from females, eventually he'll become so aroused that he'll mount and attempt sexual intercourse with any small animal or even with an inanimate object.

In the frog's native habitat, a female that is attracted and titillated by his hymeneal invariably will join him. At this, his character will undergo an instanteous and radical change from canorous troubador to mad rapist. Without any preliminaries, he will hop onto the female's back and fiercely wrap his forelimbs around her body.

Even thus securely ensconced, he's not yet home free, particularly if there are other ruttish males close by. In their disappointment in being beaten out for the prize and with an increased eagerness to mate, they will attack him even after he's mounted. Once in the saddle, a frog is not easily dislodged, and he'll defend the status quo to the very end. At times the battle will be so violent that the fought-over prize itself, the female, is injured, sometimes fatally.

Even without competition the final conquest can stretch out indefinitely, one reason being that the male does not possess a penis. This not-so-minor detail does not allow him to dictate the time of actual copulation. Since his genital structure is composed of a cloaca similar to that of his mate, he is forced to wait until she makes up her mind that the moment has arrived. Some females, for whatever reason, will delay coming to climax for some time. Under such circumstances, she will hop around carrying her mate piggyback, while in the background the other males chorus their stentorian disapproval and disappointment.

During all this, the male utilizes his resources to hurry her decision by tightening his grip around her body with sufficient pressure to literally squeeze the eggs out of her. For a fertile mating the two must synchronize their discharges, she with the eggs and he with the sperm, just as the eggs emerge from her cloaca. He'll be alerted to the imminent discharge of the eggs by her abdominal contractions and he'll time his ejaculation to coincide with her release.

After a tension-filled ride and prolonged squeezing, the couple will have reached the heights of excitation which in turn will trigger the

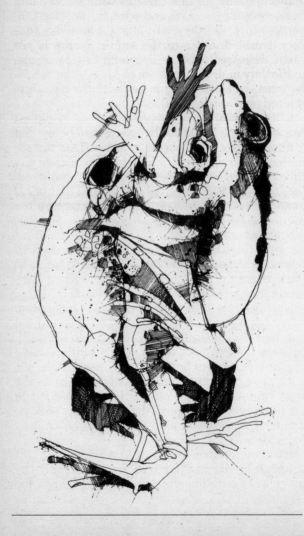

dual discharge of egg and sperm. Completely spent after the climax, both emotionally and physically, the lovers will remain in a state of quiet exhaustion for some time.

Despite their exhibitions of sexual virility and masculine chauvinism, a great many frogs are hermaphrodites when young. Usually, it will take some time before these adolescents decide which route they will follow—masculine or feminine. Most of them develop into males.

Certain species of toads never completely lose their hermaphroditic capacities. If such a male should be castrated or sustain serious injury to its testes, its rudimentary female organs will develop. Ovaries, oviducts and female sex glands will mature, and shortly he'll turn into a female capable even of reproduction without having to face emotional trauma or the need of a sex operation.

THE MOTH MITE
To Rape, Add Incest

As it may have become clear, some of the most extravagant and bizarre displays of sexual behavior are those demonstrated by various species of parasitic insects. The moth mite, which is a parasite on certain species of caterpillars, is no exception. In fact, it may be one of the most sensational sexual deviates we've had the pleasure of meeting.

During an average sexual encounter, a male moth mite, in addition to copulation, will succeed in performing midwifery, incest, infant sexual abuse and rape. These insects are viviparous (live birth) and, after birth (the males are born first), the young males hang around their mother's vagina.

Here, they feed on her juices for sustenance and wait for their sisters to be born.

When a female infant appears in their mother's birth canal, one of the newborn males grasps her with his hind legs and helps deliver her. He then immediately mates with her. The entire birth and mating operation is completed within a moment or two, during which time his brothers are engaged in identical practices.

We shouldn't be too harsh in our judgment of these parasitic insects. If the males didn't take sexual advantage of their infant sisters they could not otherwise find sexual partners and the species (valuable to humanity in that it helps control destructive caterpillars) would quickly perish. Their seemingly deviate sexual behavior is imposed upon them by necessity and the dictates of nature.

THE PARASITIC WASP
The Eldest Son Inherits
All Sexual Privileges

Another parasite, and another ally to humankind, the wasp that invades the green vegetable bug, also employs an eye-popping form of rape and incest. This insect practices a common cultural tradition found among humans in that it passes on the family inheritance to the firstborn male. In this case, however, the treasure is always composed of young female virgins.

To begin with, the sex ratio is favorable to the male, with two or three females born to each male. The males develop more quickly and emerge from the eggs before the females. The first male born immediately takes over the territory of the egg mass.

Being older, stronger and first on the scene, the

firstborn, when his brothers are hatched, drives them away. As each female emerges from an egg, he promptly mounts and mates with her. The male is kept very busy watching his sisters hatch and then copulating with them. He is indefatigable and will have sexual intercourse with all of the female hatchlings, down to the very last one.

There are species of these parasitic wasps that perform even more amazing sexual exploits, such as the group that procreates without even using males. The females lay fertilized eggs that hatch fertilized females only.

These parasitic wasps practice a wide swing of the sexual pendulum—on the one hand providing males with a surfeit of virgin females, and on the other hand eliminating the need for males altogether.

7

SADOMASOCHISM

ONE MAN'S MEAT
IS ANOTHER MAN'S BOTTOM

One of the most frightening sexual terms to con-
temporary society is "sadomasochism," the mere
mention of which conjures up visions of whips,
thigh-high boots, chains, shackles, blood and
wounded flesh. Those most frightened and dis-
gusted, naturally, are persons who have not experienced such sex acts and whose only knowledge of them is based on dirty jokes or pornographic materials.

In actual practice, most sadomasochist engagements are not cruel and do not draw blood, but
they do enhance sexual excitation. This is one of
the most misunderstood of human activities and
an impulse toward it is found to some degree in
each of us. For every practicing sadomasochist
there are hundreds who confine their feelings to
fantasies and perhaps thousands who suppress or
sublimate their feelings.

Whether active or suppressed, open or secret,
sadomasochists become sexually aroused by suf-
fering in order to obtain sexual pleasure. It's not
the actual pain itself that produces the pleasure,
but the emotions that are provoked by the pain.

Whether these needs find their sources in
childhood corporal punishments, whether their
purpose is to bring deadened feelings and mus-

cles to life, or whether they're used to arouse sexually deficient libidos to satisfactory performance levels, sadomasochism, within bounds (to be mutually decided on by the participants), is a method that produces sexual excitement and intense orgasms. Let us also recognize that the act of sexual intercourse in itself is, to some extent, sadomasochistic, as one partner submits to the other.

A common reaction of those who are unfamiliar with sadomasochism is that this behavior is "against nature" or "unnatural." Let's take a closer look at the "unnatural" aspect. It's universally acknowledged that all lower animals, without the benefits of great intellect, imagination, memory, conscience and other human attributes, and with full reliance on nature's built-in instincts, are completely natural in their sexual activities.

With this in mind, think of the times you've watched a couple of dogs making social contact. They might commit, or attempt to commit, sexual intercourse, cunnilingus, fellatio, *soixante-neuf*, exhibitionism, voyeurism, analinctus, sodomy, homosexuality, rape, urolagnia and, naturally, sadomasochism. And all this could occur while they were just greeting each other!

A safe statement would be that for most members of society, sadomasochism is shrouded in ignorance, misinformation and fear while most animals ignore it or take it in stride as a natural aspect of their sexuality.

Among domestic cats, for example, after some period of courtship with its caterwauling, copulation begins in earnest when the male bites the female on the neck. A similar technique is common among wild horses during mating when the male viciously bites the female on the neck while she often reciprocates by kicking the male on the chest. Generally the mating behavior of the turtle is leisurely and low-keyed, but on occasion an impatient male will try to accelerate the female's sexual response by snapping at her head and legs, slapping at her face with his claws and striking her cloaca with the hard tip of his tail.

Apparently, this violent behavior toward each other is a necessary ingredient for the insurance of sufficient sexual arousal to lead to successful fertilization.

All of the varied demonstrations of sadomasochism in nature have one thing in common—they get the job done. The procedure succeeds in stimulating the participants to the necessary high pitch of sexual tension so that the mating act which follows is fruitful.

THE SNAIL
Lurid, Sadomasochistic,
Bisexual Love Life

One of the lowliest creatures among us, common to every garden, would seem to be doomed, because of its seemingly physical defects, to a humdrum, barely adequate sex life. Yet this homely little denizen of our backyards leads a love life so salacious that it would seem to be straight out of Marquis de Sade.

It's the snail, or gastropod, a word that in Greek means stomach-foot. How any creature possessed of only a single sticky foot, a tiny mushy body, slow and clumsy in movement, and encumbered by an attached shell several times heavier than its entire body could excite a sexual response from even another snail is a classic example of how sexual desire prevails over all. These supersexed mollusks indulge in sexual shenanigans that, in fairness to the rest of us, should get them arrested.

When two sexually receptive Roman snails (which are the edible variety) meet on a dark night, the sedate vegetable garden becomes transformed into a veritable Roman bacchanalia. For openers, the two voluptuaries stand foot-to-foot, glued to an impassioned kiss while rocking back-and-forth in a wild, throbbing dance.

When this initial feverish state mounts to an even higher pitch, one of the snails will release its *spiculum amoris* (love dart), a chalky, spearlike substance that pierces the other's skin. The wounded snail will react with pain but, even more, with a renewed upsurge of passion as it reciprocates with its own love dart. Sadomasochism, or pain-in-pleasure, apparently is a sexual expression that some snails have in common with some humans.

Incidentally, the love dart serves not only to zoom the libido up through the ceiling but also aids in determining whether the two snails are of the same species and therefore sexually compatible. Several species of different-sized snails may inhabit the same area. Upon meeting in the dark, an amorous couple, regardless of their desires, must first determine whether they were made for each other.

The larger snail not only possesses a bigger love dart but also is much more boisterous in its amatory activities. Therefore, a snail on the receiving end of a love dart that is too forceful and overly large will promptly withdraw into its shell. Similarly, a snail will realize its gaffe if it experiences a chalky spear or arrow that lacks in size and vigor.

Returning to the love combatants who have just driven each other to the heights of sexual frenzy with their love darts, we find that each will be protruding a gigantic, erectile penis. No, they're not strictly homosexual—they're hermaphroditic. Each possesses a complete set of male and female

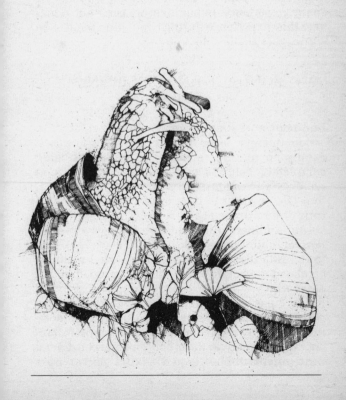

genitalia. (When reference is made of someone "having everything," snails definitely should be considered.)

At this stage, unable to contain themselves any further, the two writhing snails unite, each inserting its male member deep into the other's vagina. For several minutes the ecstatic couple remains locked in a bisexual embrace as they fertilize each other's eggs, utilizing and occupying four separate genitals in the act.

Finally, spent, they drop away from each other and lie exhausted and motionless on the ground for some time before they recover enough strength to go their separate ways. It's not known whether snails experience humanlike orgasms but whatever they experience during copulation seems like a fully acceptable substitute.

BONDAGE AND DISCIPLINE
Master-Slave Lovers

Bondage and discipline is closely related to sadomasochism even though, in some forms, it entails no physical abuse. The feeling of being completely helpless while chained or shackled removes all responsibility from the "slave" and, thus, he or she has no choice but to endure all sorts of humiliations and abuse. This condition of absolute vulnerability leads to a falling away of all normal inhibitions and leaves the "slave" exposed to the fullest range of his or her sexual feelings, escalated to heights that can be reached in no other way.

The "master," conversely, arrives at the same ecstatic goal as the "slave" even though his position is exactly opposite: feelings of absolute, unchecked power over the other. This unaccustomed position of unchallenged sexual power over another person also strips away the usual inhibitions, thus leaving both participants free to indulge and experience their sexuality to the highest possible degree.

The roles of "slave" and "master," though at opposite poles, are interchangeable, and many couples reverse them as a regular practice. Suffice it to say that the practitioners of bondage-discipline have found their sexual Nirvana.

The animals that practice their own brands of bondage and discipline find nothing unusual in their techniques; these are not only natural but the only kind of foreplay they know about prior to mating. The dragonfly, for example, utilizes bondage-discipline in its courtship rituals not from preference but of necessity.

Due to the structure of its body and the placement of its genitals, the male dragonfly during courtship never relinquishes his grasp of the female's neck, first with his feet and later with hooks on his tail. If this weren't a sufficient indignity, during part of the prenuptial flight the male persists in banging the female's head against his genitals. Instead of resenting such treatment, the female cooperates fully in the courtship; she performs a difficult maneuver by making a loop of her body that permits the mating to take place. Other animals, of course, have their own particular techniques.

THE AFRICAN CLAWED FROG
How to Manhandle a Much Larger Female

The description of the general mating habits of frogs, found elsewhere in this book, has nothing to do with bondage and discipline. On the contrary, the frogs described, particularly the males, have a direct and abrupt approach to mating, by immediately hopping on the female's back. Because of the many genera, multiplied into numerous species, not all members of the amphibians perform in this manner.

For example, in the species of the African clawed frog the male is much smaller than the female and he lacks the size and strength to bring on a climax by squeezing the eggs out of his mate. Consequently, when a male hops aboard a female, physical embrace notwithstanding, he could well be in for a ride of indeterminate duration.

When his sexual frustration grows too great, he has a spectacular, though rather chauvinistic, method for hurrying the climax. He'll dismount, grab the female and spin her around a number of times until she becomes so dizzy that she lapses into a state of semiconsciousness. This maneuver apparently has the additional effect of stimulating the female to the point where her eggs are easily released. The male also will have become highly aroused while spinning the female and he'll quickly mount his helpless mate. In this particular species, the larger and more powerful female would hardly allow her puny mate to push her around unless she were agreeable to the method and approved of its consequences.

THE LABYRINTH FISH
Male Chauvinism Under Water

The labyrinth fish is a small species. The procedure of these fish is that the dominant male places the female in a submissive role during a courtship ritual. The male, in all his resplendent colors, with spreading fins, swims around in a circle closely accompanied during the courtship swim by the female.

There's a striking difference between the two fish. The female has donned a pale, subdued color and keeps her fins drawn in tightly against her body. In addition, she swims perpendicularly, standing on her tail, parallel to her domineering mate. The female, already burdened with a bellyfull of eggs, has by far the more difficult role in this ceremonial.

THE RIVER BULLHEAD
Bullheaded and Bossy

The male bullhead runs the whole show as he takes the maternal role in various family affairs such as building the nest, watching over the eggs, and bringing up the hatchlings. This role reversal is not unusual among some species of small fish but the bullhead possesses other idiosyncrasies in his mating habits.

After scooping out a nest on the stream bottom under a rock, the male will lie inside the nest with only his head protruding from the entrance. Here he'll wait for a pregnant female to swim by, at which time he tries to seduce her by biting her on the tail. This usually has the desired effect of "turning on" the female and getting her to swim into the nest. Once inside, she'll remain a captive until she produces her spawn, which may take some time.

She spends most of her time in the cave lying on her back and frequently, the male will position himself beneath her, also resting on his back, and the pair will lie there, sometimes for hours, looking at the ceiling. Both the prone position and the duration of the spawning are unusual in that most fish make this a quick process; the female lays the eggs and the male, swimming right behind her, coats them with his sperm.

When the spawning of the bullhead finally takes place, the eggs are laid on the ceiling, also an uncommon procedure. At this point, the male acts similarly to the males of many other species and proceeds to kick the female out of the nest. For the next two weeks, until the eggs hatch, he keeps constant watch over them. During this period, he fans the water with his fins to provide aeration which increases the oxygen supply to the eggs. Following hatching, the male cares for and protects the young fish by preventing their leaving the nest prematurely.

THE DYSTICID
A Very Feisty Beetle

The dysticid beetle is an insect of diverse talents; it can swim, fly, and hang upside down for lengthy periods with its suctionlike feet gripping the surface of the water.

Most of its life is spent as a bachelor and a fierce predator to most other insects in the pond. In later life it becomes a musician, producing sounds by drawing its hard leg and wing across its scaly belly. It's this sound that attracts a mate, which is the first insect he meets that he doesn't try to eat.

Though he's able to restrain himself from gobbling her up he's not particularly tender in his romantic approach. He grabs her and with the suction cups on his feet is able to attach himself firmly to her. For the next few weeks she's his intimate prisoner as they exist as almost a single organism. They do everything in tandem—swimming, hunting, feeding and copulating. Finally, the time comes for her to lay her eggs and only then does he allow her to leave.

The larva emerges as a water tiger and the father which, till now, was one of the most feared hunters in the pond, has to flee for his life. Not only does he fail to recognize his offspring which is larger and fiercer than he, but he has to avoid the offspring in order to avoid being eaten, which undoubtedly is the highest form of disrespect that any parent could suffer.

8
SUPERSTUDS AND NYMPHOMANIACS

SEX AMONG UNEQUALS

Orgasmic ability seems to be today's most commonly accepted yardstick in measuring the maleness and femaleness of our society's inhabitants. Impotency and frigidity are the great bugaboos that frighten many of us and cause us to question our completeness as men and women.

Pornographic literature, with its emphasis on fantasy and imagination and complete disregard for reality, has distorted much of the public's view as to what is normal in sexual relations. The porno stud is always endowed with a massive, perpetually rigid organ while his female counterpart is always blessed with a divine body featuring pneumatic mammaries with buttocks to match and a constantly moist, velvety, deliciously aromatic pudendum.

Even more staggering is their demonstration of vigor, stamina and capacity for copulating hours at a time, with male orgasms coming by the dozen while the female variety is counted by the scores. All this takes place in a single sexual encounter. Then, there are the multicharacter scenarios where our hero has no problem in fully satisfying a bevy of lubricous doxies, taking them on two at a time; and the converse where the heroine is even more adept at sating the sexual desires of a dozen men, leaving them happy and surfeited.

Normal humans, when comparing themselves to these supercharged, heavy-duty copulatory machines, may feel completely inadequate as sexual beings. Fortunately, most of us recognize that unreal fantasies are the basis of this pornography, and for the most part we are content with our own sexual performances.

However, even within the normal range of human sexual experience, there are wide variations in the sexual drive among individuals. Such differences have their greatest impact in a marriage where sexual incompatibility plagues the relationship and frequently destroys it. This factor alone is a strong argument for premarital intercourse, or even better, trial marriage, which will determine the sexual compatibility of the couple.

Marriage authorities suggest to those couples faced with disparate sexual drives to effect a compromise: the one with the stronger drive consciously to reduce the sexual demand and the one with the weaker drive to make greater efforts to join in sex activities even when the desire is lacking.

Such compromise in good faith can help most marriages, but what happens when the partners find themselves at opposite ends of the sexual drive? The answer is a state of extreme unhappiness that leads to infidelity and divorce. Such conclusions are inevitable when one extreme of human sexuality can mean a half-dozen, or more, sexual contacts per day, every day, as compared to the other end of the pendulum where sex twice a year is satisfactory. Fortunately for marriages in general, most of us are clustered around the middle of the sex frequency curve, in the neighborhood of three times a week.

Most other animals don't show the wide sexual differences among the members of the same species as do humans, although there are some species where the swing is even greater among the individuals. For example, an African antelope, the kob, practices selective mating in which the females will copulate with only about three percent of the males. This means that out of a herd of 1,000 antelope, split evenly as to gender, only about fifteen males will service about 500 females, thus leaving approximately another 500 males in a state of permanent celibacy.

The handful of male kobs that receive all the female favors are the finest specimens among all the males and have won their top rank by taking over and successfully defending the herd's sexual areas. It is only in these areas that the females will engage in copulation so the male proprietors of these spaces are automatically ensured of mating with the entire female contingent of the herd.

Similar sexual behavior is practiced by the sage grouse of the northern United States, the bowerbird of New Guinea, a group of small birds called ruffs (males) and reeves (females), and a few score of other species.

All of these animals practice selective mating, or sexual selection, in which the procreative activities are performed only by the largest, strongest, handsomest and most aggressive of the males. Such top-of-the-line breeding males provide the surest and quickest way of producing the highest-quality offspring, thereby improving the stock of the group and ensuring its survival. The cooperation of the vast majority of the males which unquestioningly accept their celibate state is purely instinctive; measured by human values, it is remarkable.

This form of breeding, or sexual selection, is followed by a relatively insignificant handful of species. Most animal groups practice natural selection whereby all sexually receptive females are fair game to all males in the group. Within these groups, comprising the overwhelming majority of animal species, the individuals are quite uniform in their sexual performances, showing little of the sexual variations common to humans.

When comparing different species, however, the various animal groups demonstrate the widest sweep of sexual behavior imaginable, whether in frequency, duration or opportunities. For example, where a few seconds of copulation suffice for the entire five-year life-span of the queen bee, the tapeworm engages in a thirty-five year, nonstop sexual marathon; where a female elephant enjoys a brief sexual liaison approximately every five years, producing a single offspring in that time, a female termite during a like period is subject to frequent and ongoing copulation and produces about 55 million offspring; where a male chimpanzee completes a sexual encounter in about ten seconds, a large marine tortoise takes several days to accomplish the same mission.

The list of sexual variations among animals is endless and astonishing. With over a million different species existing on earth, one can expect over a million different sexual drives, approaches and techniques.

THE LION
*Miserable Male Chauvinism—
and the Females Love It*

The king of beasts, the lion, has long held a place of importance, if not indeed reverence, in the human consciousness. In ancient times the people of Egypt and other Arab countries included the lion among their deities as, also, have many African tribes. His noble head and level, wide-eyed gaze, framed by a magnificent mane, has made him one of the most popular symbols in heraldry, showing up on numerous coats of arms among the nobility.

While most of us are content to accept the lion's reputation for courage and fearlessness, there are some skeptics, probably including the lion himself. Ordinarily, unless goaded on by hunger, he'll give human and some other mammals a wide berth. Other predators, and nonpredators for that matter, are meaner, touchier and harder to get along with than the king of the jungle.

It's not that the lion is a faint-hearted coward. As one of the very largest of the predators, he's stronger and more dangerous than almost anything he's likely to meet. Because of his size and strength, his courage is rarely called into question.

However, in late years, another of the lion's qualities has been receiving more than a little attention, and that is his romantic nature. This aspect received front-page notoriety several years ago when an elderly lion named Frazier joined the animal community at Lion Country Safari, a commercial wild-animal park in Southern California.

The presence of Frazier led to some far-reaching changes within the lion community. Prior to his arrival, several young males had been having remarkably slight success in their repeated attempts to woo the dozen or so lionesses in the compound. Instead of having their amatory attentions reciprocated, they were coldly rejected and even mauled for their pains.

It was, therefore, hardly anticipated that an an-

cient, moth-eaten, rickety male specimen, whose best days were far behind him, would be much of an attraction for the bevy of cold, haughty and supercilious lionesses. But, amazingly, the young, nubile virgins flocked around Frazier like a pack of groupies around a rock star. The heretofore chaste lion compound turned into a churning and unceasing sexual orgy as the females lined up to await their turn for servicing by the venerable Frazier.

To have some notion of Frazier's sexual heroics, it's important to be aware of the prodigious sexual appetite possessed by the average lioness when in heat. One observer who recorded the performance of a lioness in estrus, reported that in a period of about 60 hours she copulated 170 times, which averages out to a mounting every 20 minutes. When the exhausted male (not Frazier) finally dragged his weary body away, another male took over on the still unsatiated female. On this basis, while we still can respect the king of beasts, it's time we gave more recognition to the real champ—the queen of beasts.

Everything considered, the lion is one of the more fortunate members of the animal kingdom as his life-style is truly fitting to his title of "king." Much of his time is spent in leisurely lolling about as the several females in the pride cater to his primary needs. The females bring in most of the food as they're far more efficient hunters than their lord, being fleeter of foot and probably meaner of temperament.

Even the pregnant female is a better hunter than the indolent male. After a kill, the male always is given first dibs on the fresh carcass while the females dutifully stand by until he's surfeited.

On the face of it, the lion seems to have a more salutary life-style than even that of a human monarch, which may explain why the king of beasts flourishes while his human counterpart has become an endangered species.

THE FLY
Super Sexual Athlete

There is little love lost between humankind and the million different species of insects that inhabit the earth. Possibly the one insect family treated with the greatest contempt and disgust is that of Muscidae, the fly. When we recognize that this family includes mosquitoes, gnats and midges, we feel all the more righteous in our sense of repugnance.

One reason for the antipathy on our part is the ubiquity of this insect, not to mention its high visibility. Of the 80,000 different species of flies (over 16,000 species in the U.S.), the one we hold in lowest esteem is the common housefly, *Musca domestica*, which is found all over the world around human habitation.

What we probably fail to realize is that this unsavory little creature is one of the world's greatest athletes. While the fly has only two wings (most insects have four or even more), his flight muscles are probably the most powerful tissue developed by any animal. The housefly can beat its wings 200 times a second for an hour and a half, or a total of one million beats during that time. To get some notion of this prodigious muscular achievement, raise your arms to the sides and move them up and down forty or fifty times—if you can.

The fly's relative, the mosquito, is even more impressive as it can beat its wings 600 times per second, and the tiny midges go over 1,000 beats per second. Because of these super muscles, members of the fly family can travel faster than any other insect or practically any other living thing. A horsefly, for instance, can fly around and around an automobile that's traveling forty miles per hour.

Aside from sheer speed and endurance, the fly is unsurpassed in agility. We've all seen them hover in the air and then dart off like lightning, or do a somersault and land upside down on the ceiling, and remain there (they excrete a sticky substance on the pads of their feet). Some of them can fly in any direction, including backward. Some indication of their swiftness and agility is demonstrated by some species of flies that lay their eggs in the abdomen of a bee, while the bee is in full flight!

Despite any points they may garner for their superb athletic prowess, we have ample reasons to abhor the fly. This little insect was instrumental in changing the course of world history, toppling powerful nations and empires in the wake of its destructive course.

It was the cause of the fall of Ancient Greece and some centuries later contributed to the destruction of the Roman Empire. The decline of other civilizations, including the decimation of the population of the known world during the Middle Ages, is also credited to this noxious insect. The fly accomplished all this havoc through its proclivity for carrying diseased bacteria that caused infectious plagues among a population.

The fly is a dirt-loving creature. That harmless-appearing fly buzzing around your immaculate kitchen is carrying two million bacteria, and it's not even loaded. Flies that abound in less-sanitary environments are not only happier but they carry nearly twice as many bacteria.

The advancements in medical and chemical sciences which introduced a variety of miracle drugs and pesticides have checked most of the widespread infectious plagues and eliminated others. This was accomplished by zeroing in on the particular species responsible for the great epidemic diseases such as malaria, yellow fever, typhoid, elephantiasis and sleeping sickness; and by destroying their breeding capabilities through extensive use of pesticides on the one hand and with universal inoculations and vaccinations on the other.

Of course, any relaxation of these preventive measures would soon see a return of the epidemic plagues of the Middle Ages, for the fly is a tremendous breeder. In addition to his other stellar athletic credentials, the fly is a super sexual athlete. It's a fair statement to say that the average male fly is ready, willing and able to copulate practically at any time and in any place. The female's nymphomania is barely a slight step behind the priapic male. After all, she does have to take some time out to lay her fertilized eggs.

Elsewhere in this book there's a description of swarming by mosquitoes. This mass sex orgy also is a common technique among gnats and midges. While most members of the family Muscidae don't resort to swarming, there's no end to the variety of their sexual techniques.

For example, the members of one species of gnats mate inside the ears of jackrabbits. At first blush, this practice has all the earmarks (no pun intended) of a very kinky sexual deviation, but that's not the case at all. The insides of jackrabbits' ears are where female gnats feed, and any randy male gnat knows that's the place to find them.

Robber flies are among the largest and fiercest of any species of the fly family and are voracious predators of other insects. When in the mood for love, the male flies around, humming noisily and

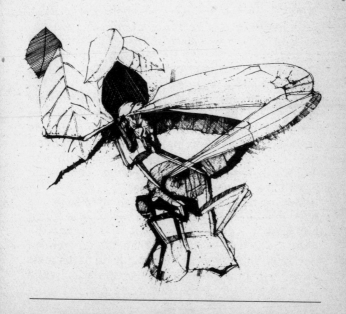

searching for a mate. He doesn't stop to flirt with just any female of the species he may meet because she's even more bloodthirsty than he is, and he could well wind up as her next meal. To most of us, contemplating an affair with someone who might cannibalize us in the process seems like an insoluble problem, but the solution is quite simple for the fly. The male robber fly crusies around until he spots a female busily feeding. That's his opportunity; he quickly alights and mates with her, knowing full well that she won't allow anything as mundane as the copulatory act to interrupt her dinner.

A similiar kind of *femme fatale* faces the male of the genus *Empis*. These are commonly known as dance flies because the females cluster in groups and move or "dance" constantly; some of them dance up and down in the same space for hours. The male of this genus is far more refined in his technique than the robber fly. Rather than save his skin by intruding upon a female engrossed in feeding, he accomplishes the same end in a far more debonair fashion, employing psychological finesse. When he comes to call on his lady, he brings her a gift.

These gifts are varied, their sophistication depending on the evolvement of the species. The more primitive male will bring his intended bride an insect or part of an insect. As she munches on the tidbit he can mate with her in safety, though quickly, before she's through eating. This might be called a hurried wedding feast.

Moving up the scale, a more advanced species will find the male carrying an insect in a filmy, transparent balloon, whose fancy trappings delight the female even more than the enwrapped delicacy. This gift wrap is a great boon to the male as it allows him a more leisurely copulation while his mate is occupied with unwrapping the gift.

The true sophisticate is the male fly that doesn't even bother with collecting an insect but merely brings his beloved an empty, though fancy and diaphanous balloon, the fine, silken threads of which were spun from his anus. She's always entranced with such a lagniappe, admiring it as she turns it over and over while her lover enjoys his sexual embrace.

These courtship customs are fascinating and give us a step-by-step pictorial of evolution's advance. One can't help but wonder what ploy the fertile instincts of the Casanova of the Empidae will conjure up next to bedazzle his tempting but deadly paramour.

Not all flies are a menace to society; in fact, some of them are of great value to us. One of these is the common fruit fly, member of the genus *Drosophila*. This creature has been a star in the experiments of medical science where it has contributed valuable information regarding the mechanics of heredity and inheritance of characteristics. One reason for the fruit fly's popularity as a laboratory subject is its high sex drive and its enthusiastic cooperation in breeding activities.

Fruit flies begin their courtship with the male tapping the female with his forelegs. The object of this is not to get her attention but to get an opportunity to taste her. His taste organs are at the tips of his feet, and this act of tasting is to ascertain whether she's of the appropriate species. If satisfied, he begins to circle her, vibrating the wing nearest her. After some circling he licks her vagina with his foot and that does it for the female. She then spreads her wings and vagina and is mounted by the male.

One species of mosquito has a most repulsive and shocking manner of mating, even for a member of the fly family. A male of this species searches for female pupae (the cocoon state before the young emerge) and on finding one will alight and slit open the protective covering with his sharp penis. He then proceeds to copulate with the not yet fully born infants. This is rape in its most reprehensible form. Its sole redeeming feature is that it's happening to a mosquito.

The common housefly, *Musca domestica*, dispenses with courtship rituals entirely. A male will approach a female, satisfy himself that she's of the appropriate species, and then abruptly hop upon her. He then strokes her head with his front legs several times and that's sufficient to remove any remaining inhibitions she may have. Copulation occurs forthwith.

The common household mosquito also ignores courtship and engages in sexual intercourse immediately upon encountering an appropriate member of the opposite sex. (Could it be something about the ambience of our households that encourages such a direct, no-nonsense approach to sex?)

As already mentioned, the fly is a prodigious sexual athlete. A pair of flies frequently copulate many times in succession, such matings being interspersed with the female's egg-laying. Each time she returns, the male mounts her and she responds by spreading her wings and extruding her vagina. This alternating of mating and egg-laying can go on for some time.

These extraordinary sexual feats only partially explain the presence of the vast number of flies. Their fecundity also lies in the brief incubation and hatching periods of their reproduction cycles. Eggs that are laid quickly hatch into maggots and metamorphose into sexually mature flies within a week or two. On this basis, within a five-month breeding season, a single fertilized fly can have descendants that number in the trillions—if they all survive. We can be thankful that they don't.

Furthermore, the fly has an astounding adaptability which gives it an enormous range of breeding places. Various species lay their eggs in hot springs while others prefer snowbanks. Every part of the earth and the ocean, both in the depths and on the surface (one long-legged species is able to run on the water's surface) is habitable to one species or another.

A most peculiar species (even for flies) frequents pools of crude petroleum. They lay their eggs in the pools of oil and use scubalike tubes that protrude above the surface to provide the pupae with air. The crude oil also provides these flies with food as they scavenge the pools of other flies and insects that have become trapped in the sticky substance.

Flies breed and hatch in an enormous number of living organisms, both animal and plant. A partial list of the animals which become unwitting hosts for the eggs or larvae of the various species of flies are spiders, rodents, monkeys, snails, caterpillars, grasshoppers, horses, cattle, sheep, deer, termites, ants, bees, and, of course, humans.

Couple their prodigal fecundity with their uncanny adaptability and it's a wonder that the earth isn't completely carpeted with a mile-deep mass of flies. Of course, nature wouldn't permit such an overwhelming monopoly and has set up a balance of predators and other unfavorable environmental conditions to keep the fly population within bounds.

Throughout history the fly has wreaked havoc not only upon humans but also upon other animals and plants, but it must be observed that only a small number of species are a menace to the rest of us. Many groups of flies actually are a boon to us as they are responsible for a most important function—the decomposition of vast quantities of waste. Further, many are predaceous and parasitic to many varieties of destructive pests.

However, the deadly and destructive species of flies must be eliminated, and science has hit upon a fiendishly clever method to accomplish this. It is focusing on an area where the fly seemingly is at its strongest—that is, sex. This is done by sterilizing the male, while allowing him to remain randy and sexually competitive. The feverish sexual activities will continue unabated, for a while. But the billions of eggs laid by the millions of females therafter will be sterile and lifeless.

An even simpler method to control these pets was used by the Los Angeles County Agriculture Department recently when the area's crops were threatened by an infestation of oriental fruit flies. These insects inject their eggs beneath the skins of fruits, which then become the food for the pupae. An unchecked invasion of these pests can wipe out entire crops.

As their first-line defense against the airborne blitz, the Agriculture Department personnel sprayed all the trees, telephone poles and street signs in the infested area with a chemical known as methyl eugenol. This chemical is a mixture of pesticide and synthetic pheremones that smells exactly like the female sexual odor used to attract the males for breeding.

Consequently, the sprayed trees, poles and street signs were attacked by hordes of concupiscent male fruit flies looking for sex but receiving, instead, lethal doses of insecticide.

These methods are ingenious in their simplicity and point up an unyielding rubric: "Live by sex—die by sex."

THE SAGE GROUSE
A Night to Remember—
A Year to Recuperate

The prize for sheer sexual capacity and endurance would unquestionably have to be awarded to the sage grouse of the northern plains of the United States. To get a measure of this bird's phenomenal sexual prowess, and using a yardstick with which we're all familiar, the sage cock could be said to average two sexual encounters a week or about 100 for the year.

That's no big deal since most of us nonchampions do as well or better than that—some of us, lots better. But the sage cock does his yearly quota of about 100 matings in a *single night!* Try topping that.

This mind-boggling sexual orgy has its beginnings about three months prior to its consummation. In the early spring, the heretofore peaceful and noncompetitive cocks sally forth to a nearby piece of ground which is to be the sexual arena for the future sexual Olympics. Here, the cocks, which are quite vain about their plumage, understandably so because their feathers are beautiful, stake out their individual territories which are about ten feet square.

For the next three months they strut around, preening, posturing, bickering and fighting, all for the purpose of improving their individual locations. That's because only about one percent of the spaces are of value; that is, only about five out of the 500 spaces have sexual significance—as we shall see further on.

At the end of this competitive stage, the desirable territories are ruled over by the largest, strongest, fiercest and most finely feathered males in the flock. In addition to the master cock, each such territory is staffed with a sub cock and several guard cocks.

Eventually, the night of nights, Orgy Night, arrives and the sober, austere prairie landscape suddenly is converted into a sybaritic, orgiastic bacchanalia as the entire female population of the flock, approximately 500 hens, descends upon the arena.

All the hens, small and drab compared to the males, try to squeeze into the five limited areas that have sexual significance. The master cocks waste no time on greetings or other preliminaries but go right to work, servicing one hen after another. After some time of steady copulation, even a master cock will slow down and finally stop from sheer exhaustion.

The as yet unserviced hens, patiently waiting their turns, are loyal to the master cock and will wait, even up to an hour or more, for him to rejuvenate his sexual powers. Even with his extraordinary virility and the patient understanding of the females, he often has to call it quits before all the females have been bred. At such time, the sub cock will eagerly take over the mating duties of his master and even some of the lowly guard cocks will manage to sneak in a few servicings.

This is another instance of sexual selection where the largest, most aggressive and handsomest males perform practically all of the mating activities. The resulting offspring are far superior to what they would have been had the breeding been available to all males.

The great majority of the cock population, reduced to spectator status at this fertility festival, must be understandably frustrated. But they understand and observe their roles as do the hens, none of which would attempt to cooperate sexually with one of these outsiders, even if he tried his utmost to woo her.

According to Webster's dictionary, the origin of the word "grouse" is unknown. I would like to submit that the name is eminently suitable. When ninety-nine percent of a male population is denied sexual satisfaction, they have plenty to "grouse" about.

THE COCKROACH
Ugly, But Oh, So Sexy

Though most people view the appearance of a cockroach in a kitchen with undisguised aversion, there are a few who must see some kind of charisma in the insect. At least, with "La Cucaracha" and archy and mehitabel, the insect has been immortalized in song and story. Also, it's been around for some 300 million years, giving it a big edge as regards First Families over the newly arrived humans who hold it in such contempt.

There are several reasons for the cockroach's longevity as a family, one of which is a fine instinct for self-preservation. They're found only where food is available, and they're not picky eaters, being interested in just about anything you have in the kitchen, so they're not likely to starve. They're mostly nocturnal, hiding out during the more dangerous daylight hours. No matter how suddenly a predator may appear, the roaches always seem to know where all the exits are for a hurried departure.

They're among the most intelligent of insects, and pesticides are not very effective in eradicating them. They simply avoid all sprayed areas until the toxic poisons have disappeared. Finally, vital to their survival, are their extremely active breeding habits; sex is a major and ongoing interest in their life-style.

While members of a species usually congregate

in numbers, thus offering continuous mating opportunities, many females and some males produce a sexual odor, or pheremone, which advertises sexual availability to anyone of the opposite sex which may be interested. And interested they are at all times.

At this time, there are over 3,500 known species of roaches and probably 3,500 different courtship and mating techniques among them. One thing they all have in common is a highly developed libido. It's probably a standoff as to whether roaches copulate to survive or survive in order to copulate.

In one common species, the German roach, a mating pair will begin its courtship with both individuals facing each other and rubbing antennae. It doesn't take much of this to get the male aroused and his next position is to wheel completely around so that his rear is facing the female. He then raises his wings in an invitation for the female to mount him.

If she, too, is stimulated, she'll climb partway up his back and feed on a special glandular secretion. A little of this is enough to give the male several erections (he has several penises, all of them hook-shaped) and he begins to back further under the female until one of his penises hooks into the female's tail. He now has her secured. He continues backing up until he's entirely out from under her body, at which time he turns around completely so that they're tail-to-tail. The remaining male copulatory members are all hooked into the female's rear end and the couple is firmly locked into a sexual union. They will remain in this position of adjoining posteriors for up to two hours of steady copulation.

The courtship ritual of another species has the male lowering and raising his body repeatedly (as though doing push-ups) in front of the female. If this has its intended effect of sexually stimulating her (or mesmerizing her?) they will proceed to mate.

The male of one species initiates the courtship by shaking his abdomen and constantly butting the female with his head. This apparently gets her into a romantic mood (or perhaps she gets tired of being butted) and is followed by copulation.

The male American roach is either more virile or more impatient than his cousins. When he gets a whiff of the female's pheremone, he dispenses with all preliminaries, heads straight for the female, and commences copulation at once.

In the more primitive species, the pregnant roach, which carries its fertilized eggs in a packet, merely drops the eggs on the floor or shelf, leaving the young (if they're fortunate enough to hatch) to fend for themselves. The more advanced species practice viviparous (live) birth. After the female drops the egg packet, she retrieves it and inserts it back into her body where the young roaches will hatch, later to emerge as living young.

As with flies and other destructive pests, scientists have hopes of controlling cockroach populations by taking advantage of their hyperactive

sexual appetites. This will be done by chemically reproducing the roach's sexual odor, the pheremone, and using it to entice and entrap the insects in large numbers. Well, at least they'll go happily.

THE KOB
The Owner of the Property
Gets the Girl

Property rights, one of the most widely held tenets in human society, is also important for many animal groups. Among some species a proprietary interest in territory is even more precious than that found among humans.

A graceful antelope of Central Africa, the kob, has developed a society which places prime importance on property values. Whereas with humans the landed gentry usually capture the most desirable women, the landed kobs *get all the females*. A large distinction is that the kob has to earn his proprietorship and then defend it every day. There are no inheritance laws or customs among the kobs.

A herd of about 1,000 kobs will set aside an area approximately the size of two adjoining football fields which then is dedicated for the exclusive purpose of propagation. In no other area will the females consent to take part in copulation nor will any male attempt such acts.

Within the sexual area are about fifteen separate spaces, each about fifty feet in diameter, with closely cropped grass resembling a putting green. Another refinement is that the greens have different values, with a premium placed upon those in the center of the area. The sexual values of the other greens become less toward the perimeter.

The center greens are presided over by the biggest and toughest of the bucks with the less-dominant males spreading out toward the edges.

Confrontations during the rutting period are constant as the nonpropertied males challenge the occupants of the fringe spaces. At the same time the proprietors along the perimeter try to better their property values by challenging the bucks in possession of more desirable or central spaces.

The possessor of a green has a tremendous advantage, much of it psychological, over his challengers. From his position of landlord, his fierceness, attitude and threats are far more vigorous than those of the challenger who is frequently beset with uncertainty. More often than not, the rival is bluffed away by the aggressive stance and threatening moves of the proprietor. As is common in nature, mere possession of the territory gives the owner a decided edge in any confrontation.

The females, upon coming in estrus, enter the sexual grounds to be serviced. These slender, graceful creatures can choose any of the sexual champs, and they make their way to the central spots first. During this, the bucks stand in the center of their greens, in a studied posture, looking off in the distance and paying no attention to the females. The does, on their part, appear to be equally disinterested in the males.

Even after a female has entered a space, she'll show no interest in the male but instead will pretend to graze on the stubble of the cropped grass. After this pretense has gone on for a bit, the tableau dissolves as the buck approaches the grazing doe and places his forefeet between her hindlegs. If the doe remains stationary, the buck will proceed to mount her. But even this critical move, the mounting, is no guarantee of success; quite often, at the last second, the doe may wiggle her haunches and take a step or two so that her suitor slides off. This maneuver may be repeated several times, at the end of which the doe will remain still so that copulation takes place.

Or the doe may lose interest and trot away to the next green. In such event, the buck will not try to prevent her leaving, chase after her, or fight the neighboring male for her favors. Instead, the rejected male ignores the departing female and resumes his haughty, disinterested stance, with head held high and his gaze directed into the distance.

From this it's clear that the rivalry among the males is not primarily sexual but instead is territorial. This is misleading because sexual rights accrue only to those who have first assumed property rights. The two rights, property and sexual, always go together.

Such sexual procedures are possible only when the highest kind of sexual morality is strictly observed by all members of the society, as with the kob. In this species, no female will copulate anywhere but in the sexual grounds; nor will any male attempt sexual intercourse outside of this area. Certainly, human sexual morality cannot begin to be compared with that of the kob.

This conduct of sexual affairs in the kob society is guaranteed to keep the adrenalin flowing in the dominant males as their days, during breeding season, are filled with confrontations and encounters. Each of them may be challenged a couple of dozen times a day for possession of the precious territory with probably a like number of sexual encounters.

Between challengers and coquettes, coups and copulations, the lord of a territory has no time or energy for anything else. Undoubtedly, he must breathe a sigh of relief when mating season finally comes to an end.

THE STONE GROUSE
Beauty Contests—Winners and Judges

Beauty contests are a familiar event in our culture, being staged in ever greater numbers throughout the world. The reason for such events is obvious: to select the most beautiful (that is, most sexually desirable) female. The winner of each contest is awarded a title and various valuable prizes. During the proceedings, the physical attributes of the lovely females on display, always shown in revealing bathing suits, are enjoyed vicariously by the viewing audience.

The winners, and some of the losers, may go on to professional entertainment careers or even, because of the wide publicity and visibility of the beauty contest, end up with highly advantageous marriages; that is, the acquisition of husbands who are wealthy and/or famous.

Animals, too, have their beauty contests and the awards for the winners are not so different from those of the human variety. The stone grouse of South America stages an annual beauty contest during its mating season. It's conducted along the lines of any typical event of its kind. All the contestants step forward, one by one, and exhibit their glorious plumage and other physical charms for the edification of the judges.

So, one asks, what else is new? Well, there are a couple of such items. For one thing, the beauty contestants are all males. For another, the judges, who happen to be all the females in the colony, make their choices right on the spot with unmistakable awards to the winner, backing up their votes with everything they've got.

You don't need a program or a master of ceremonies to spot the winners—they're the ones copulating with the judges. The losers are equally recognizable; they're on the sidelines, observing the winners cashing in their awards.

9
SEXUAL AIDS

THE MANY AVENUES OF SEX

Artificial or mechanical appliances to enhance the normal sexual experience or even to aid those afflicted with sexual malfunctions is a controversial subject in our society. Even during these sexually permissive times, a vast majority of the population equates sex aids with perversion. As a result these aids, though legal, are difficult for the average person to obtain. That's because they're advertised only in sex or underground publications for direct-mail sales or sold in pornographic shops. Other than in this miniscule number of outlets, these items are nonexistent. On the other hand, artificial aids such as crutches, heart pacers, prosthetic limbs, hearing aids and eyeglasses are commonly accepted as necessities.

Also ignored by the puritanical establishment is the employment of these artificial sex aids since the dawn of civilization, as evidenced by the various phallic models discovered among ancient Egyptian tombs. The Japanese have openly used such devices for hundreds of years and today display them in retail store windows with full public acceptance.

Animals, of course, are rarely subject to the sexual shortcomings that beset humans. Even if they were, they'd be unable to design and manufacture artificial substitutes. However, there are a few animals which, under certain circumstances, do resort to artificial aids for relief of sexual tension.

A female porcupine, for example, not having quite reached the state of estrus, but nonetheless prodded by a genital itch, will "ride" a stick or branch to obtain some sexual relief (described more fully in another chapter). She'll straddle the branch, holding one end in a paw and the other end dragging on the ground while walking on her hind legs, pressing the branch tightly against her vulva.

During this period of sexual restlessness, the male porcupine may also "ride" a branch in the same manner as the female. Usually, this will be triggered when the male catches a whiff of the female's scent on the branch.

Such ersatz means of sexual activity are rare in the animal world, yet sex aids are used by various animal species. In these cases, however, the aids have been crafted by the master designer, nature; so in the final analysis, they're really natural.

These natural aids are described in the following pages and apply mainly to those species in which the male does not possess a true penis. The absence of this primary organ in the human male would present a disastrous obstacle to overcome, but with other animals nature has devised a simple and natural way out of the dilemma.

THE OCTOPUS
Sixteen Arms
and One Substitute Penis

The octopus belongs to the class of cephalopods which includes squid, cuttlefish and other related mollusks. Cephalopod means in Greek, head-foot, which only partially describes these creatures because they also have arms or tentacles, lots of them. The squid and cuttlefish have ten and the octopus, eight, with two rows of suckers (which act like suction cups) along each tentacle. One other description of the cephalopod's physiognomy might give you a notion of its character: its mouth is located in the middle of its foot—a sort of marine walkie-talkie.

The appearance of these marine creatures is so grotesque and monstrous that they seem to be straight out of a demonic nightmare. Regardless of how they fail to meet our standards of comeliness, they are part of nature's scheme. One thing they're not is sexy-looking, and it's hard to imagine how an octopus could be attractive and seductive even to another octopus.

Even more difficult to fathom is how two heads and sixteen writhing tentacles (twenty, if you're a squid or cuttlefish) go about making love.

Here's how they do it: One of the tentacles of the male octopus is used for reproduction. It's usually longer, and in some species much longer, than the other arms. So it's not a real tentacle, but neither is it a real penis. It does not have a duct through which the sperm flows. Therefore, it's really a substitute or modified penis and is called a hectocotylus.

When a pair of octopuses meet with romance in mind they engage in a brief courting ceremony in which the male gently massages the female. This stroking has the effect of sexually stimulating both participants. This is not so different from the human experience, but with eight arms the octopus is way ahead of us as a masseur.

After some time, this stroking has its intended effect of arousing the libido of both participants, which is outwardly demonstrated by the color

changes taking place in both bodies. At this point, the male reaches into his breathing funnel with his hectocotylus and removes several cartridges of semen. He then inserts the substitute penis into his mate's mantle cavity where he deposits the semen next to her eggs. If all goes smoothly, the semen cartridges will swell until they burst and fertilize the eggs.

During this sexual maneuvering the male's sexual organ blocks the female's breathing funnel, cutting off her supply of oxygen temporarily, which causes her a high degree of discomfort. If her sexual desire is insufficiently aroused or if the male is clumsy in his technique, the female may resist him (again, not dissimilar to the human experience). At such an occurrence, a battle royal may ensue in the ocean depths.

In some species, the male sexual organ may break off during the mating and remain lodged in the female's mantle cavity. When the eggs are ready for fertilization the detached tentacle will spray them with its sperm. Following this, the female will spew out the copulatory member. In the meantime, the male has already grown a replacement. Such detached sexual arms have reportedly been observed swimming freely in the ocean, presumably on the lookout for a female octopus in heat.

THE SPIDER
Copulate and Run

The mating procedure for spiders is an intricate and risky business for the male when one realizes that the female is usually larger than the male, always hungry, and invariably cannibalistic. Each sex has its genital orifice on its underside or belly, which makes access to the female's vagina more difficult and thereby increases the hazards for the male. To top it off, the male possesses no penis. Despite these formidable hurdles, the male spider seems as eager for copulation as any normal, randy male.

When the mood to mate hits him, the male must commence to make certain preliminaries in preparation for the love tryst. First he spins a tiny web which he presses against the sexual organ on his abdomen until he squeezes a drop of semen onto the web. Then he digs his feelers, or maxillary palp, into the drop until their vessels are filled. This act triggers a sexual excitement in the male and drives him to search for a receptive female. This in itself is an extremely hazardous undertaking.

Male wolf spiders hold their sperm-laden feelers high and dance around the female, looking for all the world like fancy-footed jitterbuggers. Some males, reluctant to confront the female directly, attract the female's attention by plucking on one corner of her web until she comes over to investigate. Other males resort to bribery and attempt to buy the female's favors by bringing her a gift of a tempting insect. She rarely refuses anything edible and when she begins to eat it the male takes the opportunity for a swift mating.

Among common garden spiders the female often takes the initiative by hanging upside down from her web, invitingly presenting her abdomen to her lover. He proceeds to caress her abdomen which, in time, has the effect of hypnotizing her, thus giving the male the opportunity to mate. But he must act quickly before her trance wears off.

Some male spiders have been observed holding the female's forelegs during courtship. Contrary to its appearance, holding hands is not an expression of tender romance but is done in self-defense. So long as he holds her forelegs, he prevents her driving her poisonous claws into his body.

One of the most imaginative stratagems employed by males to save their skins while also enjoying an intimate fling with a member of the fair sex is to truss up the deadly female in a silken web. Once she's securely tied up in the gossamer strands, the male can proceed to copulate with impunity. To some, this may smack of bondage, or at least of duplicity and chauvinism, but to the involved male it spells out the double message of procreation and sweet survival.

These and many other diverse courtship rituals are time-consuming. Among some large tropical spiders these preliminaries can take up to two or three hours. The copulatory act itself is usually a brief affair with a double ojective: to copulate and depart before the female turns her lover into another high-protein meal. This is accomplished by waiting for the opportune moment when the female lapses into a passive or receptive state and the lidded vagina on her abdomen is easily accessible. At this time, the male plunges his feelers deep into her genital orifice, releasing the semen and then quickly departing. Sometimes, in his frantic haste to pull out and flee, he leaves her something to remember him by—his sex organ. It breaks off and remains in her vagina.

In fairness to the entire order of spiders, not all females are cannibals or even hard to get along with. Many female ground spiders are more than agreeable and will rear up on their hind legs so that the vagina is easily accessible and inviting to the male.

Among some species, such as the crab spider, the embrace is warm and loving, with the male hugging the female's fat abdomen with all eight of his legs, and with no threat hanging over his head that he may become the entrée at his own wedding feast.

THE DRAGONFLY
A Mating Wheel in Flight

The forbidding, frightening name "dragonfly" is only slightly more scary than the more common names "darning needle" or "sewing needle." Many children are apprehensive over the myth that his slender, darting, zooming insect is able to sew your fingers or toes together. Scientists have come much closer to the truth with the name they've bestowed upon one species of this insect, *Calopteryx virgo*, meaning "maid with the beautiful wings."

This order of insect goes back 200 or 300 million years. In earlier days it had a wingspan of nearly four feet, making it the largest insect ever to have inhabited the earth.

Today the common dragonfly has a wingspan of about four inches. Despite its primitive aerodynamic structure, it is one of the swiftest and most agile of flying creatures. Added to this agility is a pair of large eyes that contain 28,000 fixed lenses, enabling the insect to observe an area of 360 degrees in circumference. In short, there's just no way to sneak up on a dragonfly.

Though it lives for two years as a fierce and ugly aquatic grub before its glorious metamorphosis, the life-span of the dragonfly is only twelve days. Therefore, it devotes its entire adult life to a hyperactive schedule of feeding and mating.

The male begins his day about the middle of the morning and stakes out a territory about twelve feet square along the bank of a pond or stream. Within this restricted area, he conducts all of his affairs that day. This includes hunting prey, driving off male intruders, mating with most females that happen to pass through and, finally, determining where the fertilized eggs are to be deposited.

The mating process among dragonflies is one of the most complicated procedures to be found in the entire animal kingdom. Probably this is due to poor planning by nature back in the early days. Remember that the ancestors of these insects go back several hundred million years. Nature invented compensatory features, including additional sexual parts, to overcome the initial errors.

To begin with, the male has his sexual organs at the tail end of his abdomen. That in itself isn't disastrous or even unusual, except that adjacent to the sex parts is a set of grappling hooks. These hooks are necessary to grasp the female around the neck during copulation, and this automatically makes it impossible for the male to engage his genitals at the same time.

Nature compensated for this goof by providing the male with a second set of genitals at the other end of his body, on the front end of the abdomen. However, though these secondary organs consist of a penis, a sac and another set of hooks, they do not contain the vital semen. Therefore, before a male can engage in fertile copulation, he must prepare himself by curving his abdomen around so that his primary sex organ at the tail fits against the sex sac up front which he fills with sperm. After this transfer and with his tank filled, he's ready for sex.

When a female appears, his first romantic move is to grab her around the throat with his legs. If she's agreeable, they'll fly around in tandem for an enjoyable courting flight.

In the next phase of the courtship, they alight on a branch and the male loops his abdomen around so that the tail pincers fasten on her neck, thus allowing him to remove his legs from that part of her anatomy. They fly around some more in this peculiar tandem, during which the male repeatedly bangs his partner's head against his primary sexual organ.

Finally they're ready for copulation. She loops her abdomen around the forward part of his body so that her sexual aperture is adjacent to his secondary genitals. In this position, with each body looped into a semicircle, together they form an irregular circle called a "mating wheel." Flying together in this acrobatic posture, the male hooks the tail end of her abdomen, drawing it up against his secondary set of genitals, thus permitting him to insert his penis into her genital opening so that the sperm from his sac flow into her.

The male is not yet through with her since he takes most of the responsibility for the success of the reproduction process. Still holding a firm grip on his partner, he leads her to an aquatic plant and finally releases her to climb down the plant a few inches into the water to lay her eggs on a leaf or stem of the plant.

While directing this operation, if another female flies by he'll copulate with her in the same complicated fashion and direct her to lay her eggs alongside the first batch.

In some species where the male is super-conscientious, he'll climb down into the water with the female and release his grip on her only after the egg-laying process is completed.

Despite the aerial acrobatics, gyrations and contortions employed in the mating act, the dragonfly never loses his cool or becomes disoriented. The entire action, from the time he greets the female by grabbing her throat to the laying of the eggs, is encompassed entirely within the confines of his twelve-foot-square territory.

THE DOG
Impotency? What's That?

All animals are blessed by nature with the gift of propagation. Admittedly, some find the act more pleasurable than others, but several groups of male animals, including the dog, have benefited even more from nature's largess in the specially designed structure of their genitals.

The male dog is in the enviable position of never having to worry about sexual failure or impotency. His penis is composed of tissue, as is normal among all mammals including man, but also found in the dog penis is a bone which aids in the stiffening of an erection. What better sex aid could any male ask for? Other animals favored with this built-in sexual boon are bears and martens.

Another unusual aspect of the dog's genital composition is the behavior of its penis at climax. During copulation, the erectile tissue of the penis swells enormously so that following ejaculation, he is unable to withdraw his organ for some time, anywhere from a few minutes to an hour. This delayed withdrawal prevents the loss of semen from the female while also permitting (enforcing?) an intimacy and togetherness at a particularly tender moment.

MASTURBATION
Even the Animals Do It

Every female animal, on reaching estrus, desires sexual intercourse and will signal her newly-arrived-at state to appropriate males, whether by releasing a scent, by certain sounds or by visual means.

If no mate is immediately available and if the genital itch becomes persistent, the female will attempt to obtain some relief by rubbing her vulva against a tree, a stump or other object; or she will drag her genitals along the ground. A further response to the sexual demand is for the animal to lick her genitals. Many male animals also lick their genitals when in a similar state of sexual excitation.

The male subhuman primate has many more options in methods of masturbation than does his human counterpart. Whereas man is pretty well limited to the use of his hands, among most primates the male can stimulate his penis with hands, feet, mouth and tail.

Erogenous zones in some animals are found in unexpected areas. For example, some male deer use their antlers to masturbate, the same magnificent horns they use in brutal, shattering combat with other stags. During rutting season, the stag masturbates by gently rubbing the tips of his antlers to-and-fro against a bush. Within fifteen seconds this friction will produce an erection and ejaculation.

A lion, already noted elsewhere in the book for his high sex drive, has found a relatively simple way to relieve himself. He lies on his back and manipulates his penis between his two hind paws.

Mutual masturbation is common among animals and is frequently a normal part of the courtship. A pair of courting elephants titillate and stimulate each other's genitals with their trunks, using the two sensitive fingers at the trunk tips most effectively. A male elephant, left to his own resources, is able to manipulate his sex organ with his trunk. A female elephant, with her vagina out of reach of her own trunk, will engage the services of a sister elephant for mutual masturbation.

Very often, a female primate will fondle the male's penis to erection while he will reciprocate by fingering, nuzzling and licking her vagina which produces erection of the clitoris. Such stimulation will invariably lead to copulation.

The dolphin is one of the most liberated sexual creatures on earth and will partake in any sexual activity at the drop of a flipper. In contrast to most animals (with the exception of humans) it will engage in frequent copulation with no reproduction drive involved.

Dolphins are always receptive to sexual advances, regardless of sex or age. Six-week-old males are able to engage in sexual intercourse even though they don't become sexually mature until they're five years old. Homosexuality and masturbation are common pastimes for members of this genus. Their methods of masturbation are varied and often ingenious. A jet stream of water gushing into the tank is all a dolphin needs to enjoy masturbation as he positions his penis to receive the massaging effect of the onrushing water.

In one respect dolphins are superior to humans as regards deviations. While they can become sexually aroused by many stimuli, they don't become fixated on biologically inadequate objects. In short, they never become hung up on fetishes to the exclusion of valid sexual objects.

SEX IN THE ZOO
Captive Sex vs Wild Sex

Generally, the public has observed animals in captivity where their behavior is quite different from that practiced in the wild. In a natural environment the animals have freedom and many more opportunities to engage in various activities, including the avoidance of superior predators and the business of survival. Of course they enjoy the natural excitement generated by these activities.

Inside cages, with regular feedings and other routine procedures, the animals become bored and apathetic. This affects all of their behavioral patterns, particularly in regard to sex. With little else to engage its attention, the animal often resorts to masturbation to a far greater degree than it does in its natural state. Sexual aberrations, rare in the wild, become common in a zoo environment. There are many occurrences of females in heat mounting other animals, male or female. When males and females are separated, there are numerous instances of homosexuality. Male primates, without access to females, indulge in anal copulation, with the submissive male assuming the female's role of crouching or bending over.

Many animals, when isolated from their own species, attempt to copulate with different species, including humans. Sometimes the mismatch is ludicrous (though not to the frustrated animal) as in the example of the peacock which fell in love with and attempted to mate with a giant turtle. Aroused and frustrated animals even make sexual approaches to inanimate objects.

All of these deviations are rarely attempted in the native habitat. There's no need for it in the wild where mates are usually plentiful and other activities are available.

10

SEXUAL TENSION

SPOUSAL AROUSAL

The sexual tension necessary to insure fertile mating takes many different forms among the multifarious species of animal life. Some animals require precopulatory sexual excitement of the highest order while others find comparatively mild stimulation adequate to their reproductive needs. There are yet other groups that seem to function satisfactorily without any observable courtship display and seem only to require an available partner. Some of these animal orders or families arrive at sexual readiness in a short time, while others need a lengthy period to reach the same state.

With some, the courtship rites are simple and brief, but with others the courtship is lengthy and elaborate. In all cases the courtship display, maneuvers, sounds, odors, lights and other phenomena are precise to the members of the particular species. Rarely is a courtship procedure eliminated or displaced, or a new and strange ritual added.

While some of these rites are learned, the overwhelming majority of these ritualized procedures are instinctive. An animal faced with its first courtship experience will go through a lengthy and intricate ritual without a hitch.

To the casual observer these courtship ceremonies seem to have a single purpose—the stimulation of the female so that she will respond to the male's sexual demand. However, these preliminaries frequently serve a double purpose, particularly among birds, in that they definitely determine the mutual compatibility—the matching of the male's period of fertility with the receptive phase of the female. With many species these mutually fertile periods are of short duration. It's imperative for the continuation of the species that the brief and fertile time span of the male and female be synchronized.

This is especially vital among those species where the behavioral patterns of parental care follow in a precise order. Perhaps a majority of birds must pass through the entire reproductive cycle within a narrow time range. They instinctively calculate the time needed for courtship, mating, egg-laying, brooding, hatching, feeding, care and training of the young almost to the day.

As a specific example, pigeons, after mating, are found to be secreting crop-milk in preparation for feeding the young. Curiously, both male and female share this glandular faculty.

Therefore, the courtship rituals have a far deeper design than the surface appearance of macho strutting, preening and posturing for the purpose of a casual sexual encounter. In every case, it's the introduction to a chain of events critical to the continued survival of the species.

The various courtship procedures described throughout this book point up the tension involved in the behavior preliminary to the mating. These include, among many others, the violent confrontation and sexual stamina of the mink; the love dart of the snail; the ferocious attack of the crocodile; the lengthy serenade of the frog; and the group orgy of the callicebus monkey.

These courtship expressions, varied and diverse in action, are all aimed at the same objective—the arousal of sexual desire and tension in order to assure the success of a fertile union.

THE CICHLID
Don't Play with the Big Fellows
Till You're Ready

The cichlid, a tiny exotic fish, has evolved a direct tension-building courtship ceremony. With many species during the spawning season, the usually drab colors of both sexes will turn into a rainbow of vivid hues. The male at this time will scoop out a little nest at the bottom of the stream and aggressively defend this territory against all comers except pregnant females. Intruding males will either flee at his attack or show signs of submission by fading their brilliant colors and pulling in their fins; at this, the proprietor will cease his attack and permit them to withdraw peacefully.

When a receptive pregnant female appears, the courtship ceremony will be a mutual display, with both individuals showing brilliant colors and spreading fins, and swimming parallel to each other. If the female successfully passes this test, the mating will follow, as the male will permit the female to lay her eggs in his nest which he'll promptly fertilize with his sperm.

If the female flunks the courtship test by her inability to keep up her end of the pair-formation, which can happen if the male is larger, she will then flee. At this turn of events, the built-up tension in the male will explode into rage and he'll pursue and attack her, sometimes with fatal results.

The excitement engendered in such a courtship

display between equals or, in some unfortunate cases, unequals, prior to mating, is direct and obvious.

THE TEN-SPINED STICKLEBACK
The Love Bite'll Getcha

The ten-spined stickleback is another small exotic fish that builds up a full head of steam during mating season, especially so for the male, since the greatest burden for perpetuating the species falls upon his tiny head. After constructing a tunnel-like nest of water weeds, he'll be constantly occupied with repelling invasions of his territory by other males and females, even pregnant ones unless they're ready to spawn.

A receptive female will indicate her readiness to mate by not fleeing from his attack but remaining firm in her spot. This will be the message the male has been waiting for. He'll begin his courtship with a series of bunny-hop dance movements on his head, each hop leading closer to his nest. The female will follow closely but at the nest's entrance will usually hesitate, which ordinarily provokes the male into biting her.

At this, she'll swim away and he'll pursue, giving her another bite which usually has the effect of stopping her flight. The procedure will be repeated, probably a number of times, with bunny-hops and bites, thus increasing the excitement for both individuals.

After all resistance has finally evaporated, the female will enter the nest and remain resting with her head and tail protruding from both openings. The male will then apply the *coup de grace* by vibrating his cold nose against her tail. This invariably turns the trick as she'll drop her eggs and swim out. He'll immediately follow her through the nest, fertilizing the eggs as he passes over them. The female, still hanging around the nest, will be chased away as her services are no longer needed. The male will take over the entire responsibility for hatching the eggs and raising the little ones.

In a laboratory setting, with the fish in tanks for easy observation, the inherent courtship tensions can lead to strange deviations. In a tank containing a group of these sticklebacks where there is an absence of females during spawning season, some of the males will turn homosexual. Nest owners, at mating time, are all-black in color while the females are a mottled gray.

Males which have not built nests resemble females in coloring and will approach a nest in exactly the same manner as a pregnant female. Further, they'll mimic the female's reaction to the nest owner's courtship invitation down to the tiniest movement, playing out the ritual so convincingly that they even simulate the egg-laying in the nest.

This charade will collapse only at the final curtain when the male stickleback enters the nest and to his consternation finds no eggs there. We can be sure, that upon emerging, he'll chase the female-pretender away with a vengeance.

The same serious, homosexual game is played out in a tank where only females reside. Some of the females will take on the male roles of nest building and enticing a pregnant female into the nest by emulating the male's bunny-hops and tail bites. In this case, eggs will be laid but not fertilized though the male-pretender will go through the motions. These complicated, sterile, homosexual games are the result of sexual tension and the need to relieve sexual frustrations.

Such departures from sexual orthodoxy rarely occur in the native state and are only temporary in laboratory tanks, occurring only when the sexes have been separated. All homosexual behavior is abandoned instantly when members of the opposite sex are added to the group and heterosexuality reigns once again.

THE PORCUPINE
*I'll Let You Know When I'm Ready—
and Don't Hold Your Breath*

Many animals are faced with situations of sexual frustration, though usually temporarily, particularly those species whose females are receptive for brief, distantly spaced periods. This situation is common enough with humans but most of us usually take direct steps to alleviate the situation; at least, we have the available avenues to accomplish this.

Animals are far more limited in their ability to cope with sexual frustration. They're not blessed with massive intelligence, infinite ingenuity, ambivalent psyches, a myriad of man-made rules and laws, and a shadowy sense of Godlike destiny; in brief, they lack the intellectual and emotional diversity and polarity that make the human the most difficult animal on earth to get along with.

What it boils down to is that we won't tolerate frustration if we can figure out some way to remove it. In the sexual area we have done just that while most other animals have not.

The female porcupine is a good example of limited fertility and its accompanying frustration. She's in heat only once a year, for about *four hours*. Seemingly, that kind of sexual calendar would be enough to throw a female porcupine into a permanent blue funk over the fear of oversleeping or otherwise missing out on her annual allotment of sexual activities. However, that doesn't appear to happen. When that brief magic moment arrives, the lady with the barbed quills on her back seems to have an alarm go off inside, complete with bells, sirens and fireworks; she'll suddenly turn into a nymphomaniac, offering herself to any male (porcupine) that passes by. This ravenous appetite, however, will stop as suddenly as it began and she'll take a vow of celibacy for the next year.

While this copulatory fever seemingly had burst into bloom on sudden impulse, actually the female porcupine had begun to evidence sexual stirrings some months earlier. At first, she will undergo a period of restlessness interspersed with contrary displays of lassitude. She'll stand on her hind legs and sway to the beat of an invisible drummer. At other times she will lie on her back on a moonlit night, toying with a stick or branch, occupied with thoughts that only she knows.

As her hymeneal draws nearer, her edginess will increase, manifesting itself in a variety of untypical behaviorisms. She'll bite branches and treebark or purposelessly climb up and down trees, frequently emitting low whines. At this stage the biological urge is strong but she's unable, as yet, to do anything to assuage it.

As her restlessness grows, much of the urtication will be localized in her genitals and now she'll take more direct action to appease the inner clamor. She'll rub her vulva against any available object, including the dragging of her bottom along the ground. At this time, her favorite activity will be "riding" a stick or branch. Standing upright, she'll straddle the object, holding one end in a paw and with the other end dragging along the ground, will walk around pressing the branch between her legs against her vulva.

During this period, nearby males instinctively will be aware of what's taking place and some internal changes will begin to affect them. If one of them comes across a branch that had been ridden by a female (one sniff of the branch will be enough to verify this) he promptly will place it between his own hind legs and duplicate her "riding" behavior. This may not bring any relief but it may sharpen his anticipation of coming events. In the absence of such "riding" and if his inner urgings become insistent, he'll hobble around on three paws, holding his penis with the fourth.

The big day will finally arrive. When her sexual starting gun goes off the female will waste no time springing into action. She'll approach the nearest male and, without wasting a tittle of time, will sniff his genitals. The message comes through loud and clear; he'll turn and they'll touch noses. Then, retreating a few feet, both will stand upright and approach each other, again touching noses. This last nose kiss will do it. Before anyone can say "congratulations," she'll be down in the sexual crouch and he'll be on top of her.

To mount so precipitously on a back that's armored with rows of sharp, body-puncturing quills, seems like a foolhardy thing to do even for a hyperlickerish porcupine. But no casualties have ever been reported from such engagements. Very thoughtfully, the female will flatten her spines and even bring her tail up over her back to provide a cushion for the soft underbelly of her mate. They may thus remain happily united for up to five minutes. For the next several hours, the erstwhile frigid female will be more than amenable to all sexual advances.

It's quite obvious why such a lengthy period of

sexual frustration preceded the act of copulation. With her period of fertility encompassed in such a brief time span, the female cannot risk mating frivolously, whenever her fancy strikes her. Such a union would be sterile. The several months of increasing sexual tension accomplish the purpose of exciting her to the highest pitch, at which point she and her fertile eggs are ready for impregnation. One thing we can be certain of: after an entire year of sexual abstinence and several months of erotic stirrings, she's not likely to say, "Not tonight, honey. I'm too tired."

THE ROEBUCK
The Great Chase

Possibly the most impetuous pursuer in the animal world is the English roebuck as he engages in hot pursuit of the fleet-footed doe at the dead run. She's one female that will not slow down and allow herself to be overtaken; as a result, they both participate in the courtship at top speed.

This heart-pounding chase certainly provides a maximum of sexual excitation but all too frequently, when the pursuit has finally ended, they will be far too exhausted to consummate the business for which the chase was intended only as a preliminary. However, with the obligatory race out of the way and with a bit of recuperation, the main event will inevitably follow.

GETTING UPTIGHT
FOR SUCCESSFUL SEX

As has been mentioned in several instances and described numerous times, the diversity of sexual behavior is limited only to the number of species engaged in the activity.

These sexual procedures range from the queen bee's single, brief copulatory act during her lifetime to the dolphin's wide-open, uninhibited sexual play at every opportunity; from the one sexual day in an entire year for the female sheep to the female chimpanzee's approximately 100 couplings nearly every month; from the mink's eight-hour session of steady copulation to the baboon's eight-second session; from the emperor penguin's single orgasm per year to the rhinoceros's ten or so orgasms in an hour and a half.

Simply put, some animals seem to engage in sex most of the time while others experience it fleetingly and infrequently. The one thing they all have in common is the need for sexual tension for successful procreation, each species providing its own personal brand of courtship to accomplish this.

The one-celled paramecia, as noted elsewhere in the book, reproduce by division under ordinary conditions but in times of stress they will resort to conjugation, a form of copulation.

Even such microscopic creatures, on the very first rung of the evolutionary ladder, will not copulate in the absence of excitation. True, they have no courtship ceremonies, but the mere threat of unfavorable environmental conditions produces the necessary stimulation which sweeps through the colony, resulting in group-wide conjugation.

THE BEETLE
The Victor Gets the Spoils

The stag beetle possesses a magnificent set of

antlerlike appendages on its head. Like his namesake, the beetle uses them to do battle against other males of the species. The prize usually fought over is a female and the physical confrontation provokes direct and instantaneous tension. The victor's reward is immediate copulation with the waiting female while the vanquished warrior departs the scene. Frequently, one of the beetles becomes so aroused he can't wait for the end of the fight and will copulate with the female while still in the midst of the battle.

There are other species of beetles where the male is far less vigorous than the stag beetle. Sometimes, if a mating encounter goes on for too long, the sexual tension evaporates and the weary male falls backward to the ground. However, nature has provided for this eventuality and his genitals remain firmly engaged to those of the female so that copulation continues as she goes on about her business, dragging her mate along on his back, behind her.

THE RIVER CRAYFISH

With the river crayfish, copulation usually is preceded by an all-out, vicious brawl between the mates. The male is faced with a battle royal when he tries to turn the female over on her back so that he can introduce his sexual organ into her vagina. When she resists, as she normally does, he'll have a violent, knockdown, drag-out fight before he can establish sexual domination.

The female of this species, one of the earliest fighters for Women's Equality, guarantees spontaneous tension to the mating act with her bellicose aversion to being sexually dominated.

THE ROBIN

Robins hate all other birds, especially other robins, with a passion, and a pair of them will manage to tolerate each other only for the limited duration of the mating season. At its expiration, he can't wait to kick her out and she can't wait to leave.

Hate is a prime provocation for stress and will provide the unhappy couple with all the sexual tension they'll need to breed and raise unhappy progeny.

THE SALMON

Pregnant to her gills, the salmon accomplishes prodigious feats of strength and agility while fighting her way upstream, against almost insuperable obstacles, to spawn. The male, unencumbered by pregnancy, performs in a manner only slightly less remarkable than the female.

Tremendous tensions are built up during the course of this epic odyssey. Even after arriving at their destination, the males are faced with violent confrontations and battles for possession of spawning pools. These unbearable tensions are finally and mercifully dissipated during the final drama of spawning, followed by the ultimate act of release—death.

The list of animals engaged in tension-building courtship ceremonies is endless. As mentioned earlier, among all mammals only the *Homo sapiens* female is immune to the need for biological stimulation for fertile reproduction. Even she, evolution's most advanced sexual product, is cap-

tive to the need for arousal if her desire is to experience pleasure in the sexual act.

Lives there a person who really doesn't?

11

THE TENDER RITES OF COURTSHIP

There's an all too brief period in the lives of most people when the female believes that her man is the most virile, most handsome and most magnetic male in existence; and he's convinced that she's the loveliest, sexiest and most bewitching creature in the universe. This is the mystical time when the sweep of passion is at its swiftest, tenderness at its bursting fullness, and the bittersweet sensations of love are at their most ecstatic. It is the time of courtship.

This prenuptial period invariably is the setting for the great love stories of our literature with such immortal romances as *Romeo and Juliet*, *Antony and Cleopatra*, *Tristan and Isolde* and many others. The reason these stories have made such a deep and long-lasting impact is that we identify with the characters. We have experienced that emotional "high" of fulfillment when our feelings, tissues, viscera and nervous systems are saturated with soft, tender feelings of love. As the magical time of courtship begins to fade, these soaring sensations also begin to drop from their heights.

Particularly delicious during this period is the luxurious feeling of sensuality that is a constant reminder of his or her desirability, or frequently the more intense, voluptuous feelings of sexuality. Never again will the entire range of positive, unselfish feelings lie so close to the surface, so ready to share with your loved one—such feelings as tenderness, nurturing, warmth, understanding, generosity, empathy and on and on.

Many of these familiar feelings are not the sole property of humankind, but are shared by many different species of the animal world. While courtship, in its various manifestations, is nearly universal, it takes on different forms and displays among the diverse species.

Certainly, not all animals approach courtship with the caring devotion just described; some courtship customs are quite the opposite. With the praying mantis and many groups of spiders, mating preliminaries usually culminate in sudden, violent murder. Wild horses and zebras display sadomasochistic brutality in their nuptial rites. The courtship of crocodiles and lizards is nonexistent as they abruptly and unceremoniously proceed to a direct sexual attack.

The examples of antisocial courtship rituals could continue *ad infinitum* but this chapter is concerned with the tender and loving practices of courtship.

THE BOWERBIRD
Honeymoon in a Real Honeymoon Cottage

The most spectacular foliage in the animal world, as noted elsewhere, is sported by New Guinea's bird of paradise, which also is an accomplished singer, dancer and entertainer. All this beauty and talent are directed toward one goal: to favorably impress the female so that copulation may ensue and thus guarantee the perpetuation of the species.

Several species of the bird of paradise family, called bowerbirds, have developed an additional talent as architects and undoubtedly are the most skilled builders in the bird world. The males construct little houses (bowers) in the forest, not to live in or nest in, but expressly for sexual purposes. These are literally honeymoon cottages.

The structures range from very simple "huts" composed of some branches to impressive and elaborate "mansions." Some of the bowers are partitioned and contain several separate rooms. Most are tastefully decorated with varicolored leaves, flowers, berries and other bright objects. Many of the floors are carpeted with colored pebbles. One species even paints the interior by collecting berries and rubbing the berry and its juices against the walls.

Landscaping is not overlooked as each little house has an attractive front yard brightened by colorful flowers and leaves. In front of most structures is a small stage built of pebbles, and the male spends most of his day during mating season onstage. When a female goes by the male goes into his act, singing, dancing, entertaining, all in an effort to charm the female into entering his house.

Frequently a female succumbs to his blandishments and enters the bower where, away from prying eyes, the couple can gratify its passions. However, it's not uncommon for the female to enter, admire the layout and decorations, and then coolly leave with her honor intact. She may check out several pads in this fashion before she finds the house and owner that are irresistible.

While the male is responsible for building the honeymoon cottage, the female reserves the decision as to which male will father her offspring. Once pregnant, she'll spurn the glamorous house, build a nest in the forest, and take all responsibility for the laying, hatching and raising of the brood.

One species of these birds is amazingly imaginative. Instead of building houses, the males have created a dance-game which has the same goal as building a bower—romance. What these birds don't know is that we humans have played that same game for years. We call it a Maypole dance.

The male bowerbird makes a clearing in the forest about six feet square. He then sticks a tree branch into the ground at the center of the clearing. To the branch he glues enough twigs to give it the appearance of a tiny tree. Once he has set the stage, he waits for a female to pass by, whom he invites to dance with him. He leads off dancing around the tree and she follows. He quickens the pace and soon they're going around and around in a much swifter tempo. Eventually they move at top speed and the dance resembles a chase. This has the effect of exciting both dancers. Finally, when they're both overtaken by passion, they stop dancing and commence copulating.

To modify an earlier statement, our Maypole dance is quite similar except for the grand finale.

It should be noted that, like the sage grouse, the bird of paradise females will mate only with selected males—the largest, strongest, most comely. These males achieve their status by winning possession of the choicest sexually-significant territories, building the best bowers, and staging the most interesting games. The others, the vast majority of the males in the group, don't even attempt to perform these activities, let alone try to mate with the females.

Such procedures of sexual selection improve the stock of the species, and all members abide by the rules for the overall benefit of the group. This is an extremely high order of social and sexual morality. Humans, who have always prided themselves in being far above the moral level of animals, could never succeed with the process of sexual selection, even *pro bono publico*. Our morality would quickly crumble under such demands. Perhaps we should be more cautious in our pejorative use of the word "animal."

THE PHEASANT
Glad Rags and Happy Sex

Various pheasant species produce males with glorious, vividly colored plumage. It's the stunning effect of this Beau Brummell appearance that triggers the stimulation for the females. Some of these males take elaborate pains to prepare for mating. The male clears a space in the forest about twelve feet square, removing all growth and debris until the ground is clear and smooth. It must be acknowledged that this is pretty low-down labor for such a handsomely turned-out dude, but he undoubtedly considers it a labor of love.

Once his mating arena is in shape he delivers his sexual invitations, seemingly to the world at large, at the top of his voice. When there is no response he'll continue the love calls for hours. Eventually, some romantic or curious hen will appear on the scene. If she's sufficiently knocked out by his sartorial display, he'll mount her without further ado.

However, this kind of easy pickings is not common and usually much persuasion is called for. In fact, the female customarily takes a good look and then turns away. Turned on or not, no self-respecting pheasant hen will permit herself to be treated as a pushover. At this point, the cock will quickly run around in front of her and do a fashion display, not neglecting to fan out his brilliantly colored tail.

These maneuverings—she turning away and he running in front of her and displaying—may go on for some time, but eventually she'll succumb to the splendors of his gorgeous raiment and consent to copulation. This appears to be a lengthy and arduous proceeding for the male but it's well worth it; once the courtship has ended in success, the female will remain with him for several days. Finally, she has to leave him to lay her eggs, at which time the male begins anew his love calls for another female.

Incidentally, if a male of these species is too lazy or incompetent to clear a mating area, he can forget about any sexual prospects because no female will have anything to do with him.

THE RUFF
Hens Deflate Males

The ruff is a small male bird which places great significance on the annual mating ceremony by preparing for the event several months in advance. This preparation period is a stag party as all the males in the flock congregate in a separate area and stake out tiny plots, building a mound on each, that becomes the individual mating territory.

The ruffs, so named because of the ruffs or collars of vivid, multicolored feathers around their necks, spend this time in an ongoing competition as each male attempts to bedazzle and outshine his rivals by displaying his plumage to its most breathtaking brilliance. This technique involves the puffing out of the neck feathers as the ruff

bows his head, takes a deep breath, and holds the breath as long as possible. The resultant ballooning of the vividly hued feathers probably has little effect on his neighbors since they're all busily engaged in the identical pneumatic exercise.

After several months of such practice, the ruffs will have achieved the highest state of the art, just in time for the appearance of the first reeve, or female. Her entrance is the signal for all the ruffs to go into their act, and in unison they all bow their heads, inhale deeply, blow up their neck feathers, and freeze in the posture.

The hen, unconcerned with any male discomfort, serenely walks about, inspecting the various collars for one to strike her fancy, a procedure that may go on for some time. When she is finally turned on by a ruff, she pecks at his neck feathers.

In a flash he's upon her, copulating like a true winner while the remainder of the males lift their heads and simultaneously release a giant breath, mixed in equal parts with relief and disappointment. This breathtaking procedure is followed repeatedly until all of the flock's females have been serviced.

THE PENGUIN
Sex on the Rocks

The top prize for love, affection and unswerving responsibility to the family would have to be awarded to the penguin. During courtship, a pair of penguins displays a devotion and tenderness that would provoke the envy of lovesick human adolescents. Their considerable hugging and kissing would seem to be handicapped by the lack of arms and soft lips, but not so. In their kissing, they rub beaks and the female inserts her bill inside the

male's, while he hugs her tenderly and lovingly with his flippers. All this kissing and hugging lead inevitably to the act of copulation which, in itself, is quite brief.

After the eggs are laid, their incubation is enough to test the mettle of even the most heroic emperor penguin, which must keep the eggs from freezing in the Antarctic winter with its blizzards and temperatures that fall to 60° below zero.

During this frigid period the penguin stands unmoving on the icy snow because the eggs are perched on his feet to keep them from direct contact with the frozen ground. He holds this incredibly punishing stance for two solid months!

During this time, his mate is away at sea on a feeding spree. Upon her return, the mates exchange roles and the male finally is able to break his fast while the eggs hatch on the feet of the female where the infants rest until the weather becomes warmer.

THE PIGEON
Open Marriage

Pigeons are among the more sexually liberated bird groups with a courtship that is a tender and romantic affair, usually culminating in a long-lasting pair-bond.

Despite this base of monogomy, the female is notoriously promiscuous, which is why she's invariably followed and pecked at by her mate. He doesn't trust her out of his sight and for good reason. Practically every unattended female pigeon during mating season is accosted by every loose male in the vicinity.

These feathered Casanovas are quite casual and unmacho in their approach. If the female is uninterested the matter ends there. If, on the other hand, the female is agreeable and accepts the invitation, she'll stop and assume the sexual crouch. At this, the ball is in the male's court and he's faced with making good his promise which, frequently and to his embarrassment, he's unable to do (quite reminiscent of the human condition).

The pigeons' free and easy approach to sex is sometimes evidenced by their sexual roles when the female mounts the crouching male. The human observer rarely is aware of such role reversal since both sexes look alike even to their genitals which are the typical cloacae.

Homosexuality is common among pigeons, more frequently with males, and a permanent pair-bond of the same sex frequently occurs even when members of the opposite sex abound.

THE RAVEN
Odd Couples and Triples

When there's a scarcity or absence of males among ravens during mating season, two females will form a pair-bond and court each other, with the dominant female assuming the male role including mounting the submissive bird. This behavior carries on even to nest building and laying of eggs which, of course, are sterile.

The affection displayed by two females often is as tender and loving as that shown by a heterosexual couple.

This state of homosexuality takes on further deviations when a male appears. He often will pair with a dominant female. While finding this new relationship agreeable, this female refuses to give up her first love. This distresses the male, who tries either to ignore or drive away the submissive female, usually with no success. Though unhappy at having to share his spouse, the male finally resigns himself to the *ménage à trois*, and even breeds with both females, thus siring two broods.

The rank or pecking order is of great importance in this species, even to the roles played in a pair-formation. When a dominant female, high in the ranking order of the group, pairs with a submissive, low-ranking male, she will adopt male behavior even to taking the superior position during copulation. In such a relationship the male accepts the submissive role along with other female behavioral patterns.

Jackdaws, which demonstrate courting and mating behavior similar to that of the ravens, lead an even more affectionate relationship. Young jackdaws fall in love at a very young age, several years before they're sexually mature, and treat each other in classic "lovebird" fashion.

They spend every second of their time together, rarely separated by more than a foot or two. The male lovingly gives his bride every delicacy he finds and she accepts it in the same loving way, as they coo to each other in baby tones (reminiscent of the baby talk used by human lovers).

THE GOOSE
Till Death Do Us Part

The courtship display of geese, in terms of demonstrative tenderness and affection, is remarkably similar to that of humans. A major distinction, however, is that the goose's overt displays of love do not cease once the courtship is over but instead continue throughout their life together, which can go on for as long as fifty years. If anything, their relationship becomes more loving as the years roll on.

This is particularly noticeable with the graylag or wild gray goose. The beginning of lifelong fidelity in this species begins at a very tender age, during the first or second year of their lives. At this time, though sexually immature, the gander begins the search for a lifelong companion.

If he spots a young goose that turns him on, he hangs around the family, but not too close because the young female's father abruptly chases him away. The young male remains within view of his beloved and goes through all sorts of maneuvers to impress her. First are the long, soulful looks interspersed with demonstrations of his masculinity, courage and derring-do; parading and strutting with exaggerated steps, dipping his head underwater and, most impressive of all, instant attacks, accompanied by loud shrieks, upon any possible rival that approaches.

This combination of tenderness, aggressiveness and determination invariably pays off as the nubile object of his affection falls in love with this bold suitor. She signifies this by running away from her family to join her young mate; at this, the family accepts him. In courtship, the male is aggressive and the female is demure. If he's successful in fighting off all other challengers for her favors, he can be quite certain of ultimate success.

If it's the female that's struck by the lovebug, her approach is quite different and far more subtle. She doesn't openly chase him but just *happens* to be nearby many times during the day. After a while, he can't avoid noticing her and in time is likely to become interested.

If a drake or goose happens to be smitten with an already mated individual, any efforts to make romantic contact are in vain. Happy goose matings simply are not broken up by intruders. During their lengthy union, the drake and goose rarely undergo long separations. When together they are continually cackling love messages to each other.

This demonstrativeness reaches its peak in the triumph ceremony which is a combination of exaggerated strutting and swaggering, with wings spread, chest puffed out, long stride and appropriate calls of triumph. This ritual ends up with both mates cackling to each other, beak-to-beak. The triumph ceremony reaches its zenith when one of the couple returns after an absence. To the uninitiated, this high voltage greeting would seem to indicate that the lovers had been separated for months when, in fact, they probably had been apart for several hours.

One of the saddest creatures in the world is a drake or goose recently bereaved by the loss or disappearance of its mate. The remaining individual searches for its mate night and day, flying long distances in every direction. When it fails to find its mate it sinks into a deep depression and acquires a permanent look of sadness, especially around the eyes. Its lack of a partner to love and share the triumph ceremony destroys its spirit and its standing in the flock. Its status plummets to the bottom of the ranking order.

Such a survivor will usually take up another mate though the longer the first marriage has lasted, the longer it will take him or her to establish another relationship. In subsequent matings, the love and devotion is not as strong and, frequently, the drake will carry on some liaison on the side. Even then he's very discreet about his adultery and carries on the assignations away from his mate, usually behind some bushes.

Sometimes a surviving male will establish a homosexual relationship with another drake since it's not uncommon for two ganders to fall in love and set up a lifelong pairing. It flourishes as well as the normal heterosexual marriage except for the sterile sexual unions.

If a goose falls in love with one of the males in such a pair, she follows them around even though she's usually ignored by both ganders. If she's persistent, her opportunity to make an impression upon her loved one is bound to arise. This occurs when the gander is unsuccessful in an attempt to copulate with his buddy. Since the female is right at hand, often the drake will mount her in a successful and fertile mating.

After some such sexual incidents, she'll be accepted by her friend, though his primary attachment remains with the other drake. In time she builds a nest and hatches a brood of goslings. Both of the drakes will adopt her and her brood. Frequently, this is enough to get the second drake sufficiently interested to copulate with her and a ménage à trois is established.

With this setup, the three produce double broods and the family rises to the top of the flock's pecking order. The three combined are stronger, engage in more triumph ceremonies, and are more productive than the normal two-adult family.

THE HUMMINGBIRD
Erotic Stunt Flyers

The hummingbird, native only to the Western hemisphere, includes among its various species the smallest birds in the world; their bodies are two inches long and their eggs measure less than one third of an inch in length.

Many species have brilliant, iridescent plumage, crests and long tails; they resemble jeweled bullets as they shoot through the air. Their tiny wings beat so rapidly they're practically invisible and the birds are a darting blur as they fly in any direction including backwards. In addition, they have the ability to hover in midair which they do while feeding.

Their courtship ceremony, over which the female presides, resembles an aerial show. A group of males performs a dazzling array of flying maneuvers, each trying to outperform the others for the benefit of a single spectator, the desired female, who is calmly perched on a branch watching the aerial display.

When all of the flying daredevils have completed their repertoires of aerobatics, they congregate in front of the female, hovering until she makes her choice. The winner immediately claims the grand prize by copulating with her while the losers dart away in search of other stunt-flying courtship competitions.

THE BUTTERFLY
Beautiful, Fragrant and Sexy

Another creature which employs flying maneuvers as part of its courtship rituals is the most beautiful insect of all—the butterfly. However, these delicate varihued Lepidoptera do not enter a flying circus competition for the female; instead, it's usually a one-on-one encounter.

The male, which has powerful organs of smell on the antennae, picks up the pheremone or sexual odor emitted by the receptive female, even from a considerable distance, and flies directly to her.

The courtship begins with a pursuit, but if the female is ready for mating the pursuit is brief as she alights on a twig. The male then executes a series of graceful flying stunts before directly approaching her. This has the effect of further stimulating both individuals.

Once beside her, the male goes through additional complicated procedures. He faces the female and raises and lowers his closed wings, simultaneously moving his antennae in a circle. He then encloses the tips of the female's antennae within his wings which brings her scent detectors in contact with his scent patches. One sniff assures the female that he's of the appropriate species; each species has its own special fragrance which sexually excites only the members of that species. This is the final act of stimulation before the male mounts the willing female and proceeds to copulate.

The butterfly presents several unusual contradictions beginning with its amazing metamorphosis from repulsive caterpillar to exquisite flying creature. The most defenseless of insects, in many cases it relies on its colorful beauty for protection. Many of the most splendidly hued butterflies are the most foul-tasting, and birds, after one sampling, give them a wide berth. The drab, single-colored varieties, and moths, blend into their surroundings so that they resemble twigs on a bush or tree and thus go unnoticed by predators.

An unusual feature of some female butterflies is the complicated structure of the reproductive system. They possess two sexual openings—one for copulation and another for laying eggs, called the ovipositor. In this respect, this small, slender insect has a more sophisticated sexual structure than any other animal, including the human female.

THE FIDDLER CRAB
Colorful Beach Boys

One of the most favorable environments for a romantically inclined young male to meet a young lady of like mind is on the sunny sands of a beach. That's exactly where the male fiddler crab hangs out, with not much more on his mind than meeting females. He evinces his desire by brandishing a large brilliantly colored claw in the air. In fact, except for brief interruptions for feeding and mating, most of his day is spent in waving his claw.

The genders of the fiddler crab are easily distinguishable as the female is smaller, with symmetrical claws, and colored in drab brownish-gray while the male has asymmetrical claws (the one he waves is much larger than the other) and overall coloring of several vivid hues. Upon awakening in the morning and when frightened, the male adopts the same flat hues as the female. When the sun is up, the tide is out, and the beach is dry, the male rapidly changes to a beautiful medley of dazzling colors.

These colors announce that the crab is in a mood for romance. When an attentive female responds to his invitation, the sex hormones in his bloodstream cause his colors to become even more spectacular. The drab little females find these undulating, multihued pincers irresistible. When the females show up, the males explode in a frenzy of claw waving and frenetic dancing. We all know how nimble a two-legged dancer can be but these crabs, using all eight of their legs, are veritable whirling dervishes as they execute variations of ballet, jitterbug, rhumba and samba. In one particular dance, the female joins in and is passed from one male dancer to another.

The objective of this terpsichorean display, to

no one's surprise, is to rouse the female's sexual desire to the mating pitch. When the two reach this point, they sensuously stroke each other's legs. Then the male heads for his beach house, which is a hole in the sand. The female follows him down this hole. A moment later, the male appears at the entrance with a gob of mud in his claw which he uses to plug up the opening.

At last, they're alone.

THE TURTLE
Slow and Easy on the Draw

The turtle, despite its hard, unprepossessing exterior, is a tender romantic lover. When a pair of mud turtles meet during mating season they swim slowly toward each other with outstretched heads until they nearly touch. This is followed by a swimming ballet with forelegs fluttering and heads alternately extending and retracting. As the excitement mounts, the dance movements become more rapid until the female, overcome with desire, sinks to the bottom.

Here she prepares for her lover by moving her tail to one side and thrusting her rear end out of the shell as far as possible so that her cloacal slit is readily accessible. The male, which has followed her to the bottom, mounts her and, using his tail to guide his penis to the target, soon penetrates her. Thus united, the contented pair may be joined in bliss for hours.

The mating act of the large marine terrapin is even more prolonged as the male, after mounting and intromission, may ride the willing female for several days. Not only does she have to be willing but she also needs a strong back. Some of the larger sea turtles are nine feet long and weigh over 1,000 pounds.

This romantic interlude is extended even further if the female is not ready or if jealous rivals try to dislodge the male, even by turning the mated couple onto their backs. Such interference could delay consummation for a number of days.

As is not uncommon in many species, a male's eagerness for sex may coincide with the female's disinterest. The male then may try to get her to change her mind, having several options in this effort. In some species, the male mounts and then whips the hard tip of his tail against the female's cloaca. Frequently, this stimulates her and results in her protruding her rear end out of the shell.

If this doesn't do the trick he may snap at her head and legs at which she automatically withdraws into the shell for safety and seclusion. Since she's abnormally fat with her store of eggs during mating season, her rear will inevitably stick out of the shell with, perhaps to her astonishment, predictable results.

On the surface it would appear that turtles manage their reproductive affairs in satisfactory fashion without much need for sexual excitation. This is misleading, for without sexual tension the male would not be so insistent, slow and patient though he is, on copulating. The female, lacking a powerful inner urging, would refuse to come out of her shell and carry her heavyweight mate around on her back for hours or even days. Their lack of a highly visible sexual excitation apparently is balanced with a low-level tension that builds over a lengthy period.

As humans, we're fortunate in that we're not locked into a minklike frenzy of courtship-mating nor into the turtle-like approach of low-keyed, leisurely copulation. Instead, depending upon our moods and desires, we can enjoy either or both of those styles or something in between. Perhaps we can learn something of value from the turtle's slow-paced life-style with its languid crawl and leisurely love-making, for it lives longer than any other animal. The Mauritius turtle has a life-span of 150 years.

Their unhurried, unruffled approach to living may also have contributed to their longevity as an order; some species have survived pretty much in their original form from the age of the giant saurians, nearly 100 million years ago.

12

LOVE AND DEATH

As humans, we're all aware of the part sex plays in the creation of new life, as well as knowing the inevitability of death. Animals hold no such conscious knowledge but it doesn't matter since they, and we, are bound by the immutable laws of nature in such matters.

Paradoxically, as mentioned earlier, a form of immortality is experienced by most primitive organisms, such as the one-celled amoeba, that reproduce by subdivision while mortality is the lot of most other animals. In one sense we do possess immortality in that our living seeds and eggs form the beginning cells of new life and this germ plasma is passed on through individuals to succeeding generations. This knowledge will hardly thrill many of us since we have our entire investment and self-recognition in our bodies, or somas, which are exceedingly mortal.

Most animals do not associate the sexual act with death since the two are separated by lifespans of varying lengths. These range from the twenty-second life cycle of the bacteria to the 150 years of existence of the Mauritius turtle. However, there are some life forms, mostly among insects, where the act of procreation is closely associated with death, invariably for the male.

We're familiar with the life-and-death-style of the Mayfly as described earlier. This beautiful insect spends its metamorphosed life on earth in a one-afternoon binge of dancing and copulating. As twilight descends, the females skim over the surface of the water like tiny bombers, expelling their fertilized eggs from twin tubes or ovipositors, after which their wings drop off and they fall in the water to die.

Male bees, or drones, are bred specially as sexual consorts of the queen bee, with several hundred of them used in each mating. Though only one of them succeeds in impregnating the queen and dies immediately afterwards as he falls to the ground, the entire retinue of males is doomed as they are deliberately starved to death once the nuptial rite is over.

Another illustration familiar to us is that of the salmon which, upon reaching sexual maturity, heads back to its birthplace to spawn. During the lengthy migration to the spawning grounds, the salmon never eats again as its digestive tract has collapsed. The end of the journey is foreordained with the creation of life in the act of spawning, followed shortly by death.

The uncanny ability of the mature salmon to navigate the untracked ocean depths to the correct river and finally to the headwaters of the mountain stream where it was hatched must rival the marvelous feats of our own spacemen in their journeys to the moon. The salmon accomplishes its odyssey by memorizing the various odors of the chemical and mineral content of the waters traversed in its outgoing journey.

Other fish that follow these procedures include the sturgeon and lamprey. If it seems that nature is engaging in an unprofitable tradeoff by bartering life for death on an even basis, it's only because one is unaware of the large number of eggs laid by these fish. A single sturgeon can lay as many as six million eggs. A giant sturgeon does even better.

THE EEL
Incredible Navigators

A life-and-death journey, similar to that of the salmon, is conducted by the eels of Western Europe and the United States as they reach sexual maturity and return to their source to spawn and die. The only observable difference between the American and European eel is that the latter has a few more vertebrae. Eels reverse the process of the salmon. These odd fish leave the lakes, rivers and streams and migrate to the depths of the Sargasso Sea in the Atlantic Ocean, north of Bermuda, for their spawning. It is a miracle of navigation as the eels retrace the route through the ocean depths to their birthplace.

An even more incredible feat of navigation is that performed by the baby eels which unerringly find their way from the middle of the Atlantic Ocean to the precise shores of the inland rivers and lakes previously inhabited by their parents. This route is one they had traversed in reverse only as unfertilized eggs in the bellies of their mothers! The journey of the newborn eels is a leisurely one, as one group takes a year to reach the United States and the others travel for two and a half years before arriving at their European destinations. Though tiny larvae at the beginning of their journey, they never make the mistake of going in the wrong direction. Some years later, having attained full maturity, both branches of the family will meet once again in the Sargasso Sea, this time to create new life after which they will all die.

THE FIREFLY
Blink Once for Love,
Twice for Death

A family of small beetles, Lampyridae, have built-in beacons which they flash on and off in their systems of communications. We know them as fireflies or lightning bugs or even glowworms, which are the luminescent larvae.

Each of the numerous species within the various genera of this family has its own specific code of signals, different from all others. Some hold a steady beam for a definite time period while others blink on and off in precise intervals for a set number of times, each in its own wavelength.

In most species, the light rays emanate from the abdomen, but some are located on the sides of the insects or on the head or tail. Further distinctions are found in the colors of the beams such as greenish, yellowish, bluish or orange, or combinations of several colors. Some of the more completely equipped fireflies resemble miniature 747s in night flight.

The object of this complex system of variegated colored light signals is to get the male and female of the same species together for a proper mating matchup: a slight error in the signal, sending or receiving, would result in a wasted effort for both participants because members of different species will not mate. The precise clockwork with which these insects operate rarely allows for such errors.

Since fireflies live for only a few days, they spend the time, mostly at night, propagating. That's what their light shows are all about—propagation, pure and simple.

Among many species, the female will lie on the grass or on a bush, belly up, so that her sending apparatus is aimed in the right direction— upwards. The males fly around, emitting signals and looking for appropriate responses. When a female picks up the correct code in the night sky she responds with an answering signal. Before the male lands beside her, they will exchange signals again, perhaps several times, as a test. When he's fully satisfied that he's made contact with the right female, he comes in for a landing followed by immediate mating.

In some species, the female emits a steady glow and the male, after ascertaining the correctness of the lady's pedigree through exchange of signals, flies in on her beam for the rendezvous.

These flying light shows, entertaining as they are, harbor in their midst some not so harmless females, which are not only predatory but also are quite a bit smarter than most beetles. These females have succeeded in breaking the signal codes of other species and respond appropriately to various male signals, even those outside their usual instinctive responses. When a male of a strange species, in answer to her signal, comes down beside one of them, eagerly anticipating love and romance, he's in for an extremely rude shock. The female, invariably larger, grabs the un- suspecting victim and begins to devour him.

Some of the more accomplished of these female cannibals have learned the signals of four or more other species. Presumably they're the obese fireflies you may see waddling around. Certainly, this ability to mimic a strange light code, and then capitalize on it in premeditated predatory fashion, indicates the possession of a fairly complicated mind by these tiny creatures.

While these bloodthirsty female beetles have no problem in keeping their larders replenished, they're certainly a menace to the existence of other species. This may be an example of "survival of the fittest" whereby one species prospers at the expense of another.

One area in which all fireflies show the greatest expertise is in the production of efficient light. Their glands are able to produce a "cold" light that utilizes 98 percent of their energy for the light with only a tiny 2 percent lost as heat. Compared to fireflies, our production of light is primitive and in the class of rubbing two sticks together to make fire. Perhaps there's something we can learn from these beetles.

Most of us have seen fireflies twinkling during a summer evening in a garden or park and may have been struck by the ephemeral and ethereal nature of the lights. They disappear as suddenly as they appear, leaving nothing behind to indicate their continuing presence. However, there is a firefly in the Caribbean area which produces a luminescence of far greater intensity and longevity. It is so luminous that the natives delight in wrapping them in bits of gauze and wearing them in the hair for evening strolls when these brilliant insects serve as human headlights.

THE SPIDER
*Love in the Boudoir,
or Is It Abattoir?*

Common lore has it that the typical female spider is always hungry after copulating and that her favorite postcoital snack is her lover. There's much truth to this as the female, after having her sexual appetite appeased and becoming pregnant, has a great desire, if not need, for proteins. If her lover, usually smaller than she, remains in the vicinity she'll very likely make a meal of him.

The male, knowing of her predilection for a high-protein diet at this point, will make himself scarce in a hurry (as described earlier in the book). There are some species, though, where the male is not so fortunate.

A well-known example of a female who intends to possess her paramour body and soul is the black widow spider. This female is six or seven times larger than the male. During the courtship she's tender and loving and as sensuously feminine as one could expect of a black widow spider. Immediately following copulation she grabs her little mate for a post-lovemaking buffet, thus gratifying her two major appetites.

The little fellow is fully aware of the hazards of this mating game. Surely, even as part of him is enjoying the copulatory act, another part of him must be planning the getaway. To save his neck it's imperative that he "come" and go almost simultaneously. The sad truth is that more often than not he doesn't make it. There have been no reports on the incidence of impotent spiders but if there are such individuals, at least we'll know why.

One of the most dramatic examples of this procreation-death syndrome is found in the tropical wheel-web spiders, genus Nephila. The female is by far the larger gender, sometimes weighing up to 600 times more than the dwarf males. Fortunately, the female is not predaceous to her tiny mate, otherwise the mating game would be over before it could begin. With this in mind, the males choose to dwell in the safest refuge in the world—on the huge abdomen of the female. Thus secure and unmolested, the male lives as a parasite until he decides to mate with his gigantic sex partner.

Once they mate, it's curtains for the little males. The mating act uses up all the life resources of the tiny creature, and immediately after impregnating the female he expires. Only then does the female eat her partner.

Spiders are a unique order of insects, particularly in the sexual sphere. The male's copulatory members are completely separate from its reproductive system, thus necessitating the complicated procedures of transferring the sperm from its receptacles to the female's ovaries via the male's maxillary palp or copulatory organs, as noted elsewhere in the book. The male may not have a true penis, but at least he possesses two copulatory organs.

Spiders are not gregarious, so they have special problems when the mating urge is upon them, which occurs when they reach maturity. At this time, the male's chief goals are to fill his palpi with sperm and then locate an appropriate female in which to deposit it.

Since the female doesn't know who he is, as she's never seen an adult spider and has spent her life either in eating smaller insects or fleeing from larger ones, it's an awkward, even risky moment when the amatory male first approaches her. Her first instinct, since he's usually smaller, is to attack. It behooves him quickly to shift her appetitive craving from food to sex; this is when his instinctive courtship ritual (as described elsewhere) takes over.

Males of some species mature more quickly than the females. When this is the case, the male often moves into the female's web and waits until she attains sexual maturity. After their sexual appetites have been assuaged, their only social bond is broken and the smaller male quickly departs before the pregnant, now hungry, female has a chance to devour him.

In any event, even the males which escape are only living on borrowed time. In the act of copulation they really "give their all" as they almost completely exhaust their store of energy. From that time on, they do not eat and so manage to survive only for a week or two before succumbing.

Females of many species suffer the same fate as the males and die soon after laying their eggs. Other females, among the more than 50,000 species now known, live on for some years with annual periods of copulation. Some large female tarantulas are known to live for as long as fifteen years, during which time many suitors enjoy their favors and pay the price with certain death. All this for the mysterious and irresistible drive called sex.

THE SCORPION
The Honeymoon
Chamber of Horrors

Scorpions are members of the same class, Arachnida, as spiders, and the females are guilty of the same atrocious table manners in the way they turn their mates into meals.

Among the hundreds of species in the order, ranging from one-half inch to seven inches in size, there's a fairly constant similarity in appearance, including the appearance of menace. This is abetted by ferocious-appearing front pincers and a tail equipped with a poisonous sting. In between the two ends are a dozen abdominal and tail segments, four pairs of walking legs and three to five pairs of eyes.

Scorpions are distinctly antisocial and are content to keep to themselves. If a meeting between two of them is unvoidable the result usually is an engagement—fighting or copulating. If a fight, it's to the death with the winner devouring the loser.

Copulation occurs if the two are of the same species, of different genders and sexually mature.

The scorpion's main diet is insects. It has a precise and businesslike way of disposing of its prey. After grabbing the next meal with its pincers, the scorpion paralyzes the victim with its poisonous sting. It then mashes the insect with a pair of small pincers and injects it with digestive enzymes. The enzymes turn the prey's tissues into fluids preparatory to the final step of pumping the insect dry, leaving only the outer covering. This description of a scorpion's method of food preparation and ingestion can serve as a picture of the fate of most of the males of the order.

The scorpion is not gluttonous, possesses a small appetite, and can go without food for up to a year. However, when two scorpions do meet for courtship and mating, their heretofore solitary and reclusive behavior becomes transformed into a loving and idyllic relationship filled with graceful acrobatics, joyous dancing and sensual touching. They begin by standing on their heads, or thoraxes, and intertwining their tails in a lover's knot while using all their legs (sixteen of them combined) to stroke each other's body.

Upon righting themselves, they rear up on their hind legs and the remaining six pairs of legs (three pairs on each) clasp each other as the couple glides into a scorpion-type waltz, the male leading, though dancing backwards, and the female following in perfect rhythm. The courtship dance continues until the couple becomes sufficiently stimulated sexually, whereupon they find a cave or hole and retire to consummate their brief betrothal in private. This is one honeymoon night where only one of the lovers survives the marriage bed and emerges in the morning; the survivor is always the female. If one cares to investigate the honeymoon cave, all that will be found of the bridegroom are his partial remains.

That night, which begins on such a high note of bliss and ecstasy, alters its course abruptly upon completion of intercourse. The female, now pregnant, instantly changes from rapturous to ravenous, turning the honeymoon boudoir into a chamber of horrors as she murders and then devours her mate.

Though a savage and deadly paramour, the female scorpion is a caring and devoted mother. She delivers several score live but helpless infants. Until ready to fend for themselves, these young scorpions climb aboard their mother, frequently covering her entire dorsal area while she goes about her business, patiently carrying her entire family on her back.

THE PRAYING MANTIS
Please, Not While I'm Copulating

One of the most paradoxical of insects is the praying mantis, which conceals a rapacious character with a surface saintliness. Its bearing of quiet dignity while holding its forelimbs upwards in the traditional posture of devout supplication has led to countless superstitions and legends since antiquity.

The ancient Greeks endowed the insect with supernatural powers, while the early Moslems believed that it prayed constantly while always facing toward Mecca. Despite such beliefs, the "saintly" forelimbs of the mantis are among the most murderous natural weapons to be found. These limbs have sharp, serrated inner edges, adapted to hold and slice when the limbs fold into the body like the blade of a jackknife.

These rapacious insects prey upon flies, grass-hoppers and caterpillars, while the larger species attack small frogs, lizards and birds. They're fearless and aggressive, attacking anything within reasonable size, including other mantises; the larger eat the smaller.

Adding to their menace is their procryptic abilities—the talent to conceal themselves from predators and prey alike—as they mimic the coloration and shape of green foliage, and even flowers and blossoms, including a gentle swaying by the insect to simulate the effect of a light breeze.

The female of this bloodthirstiest of all insects is even more voracious than the black widow spider or the scorpion because she tries to consume her mate before, during or after copulation. The poor male knows that she's the ultimate female chauvinist, but when she's sexually receptive he finds her irresistible.

It appears to be an impossible situation but the male has one slight advantage which he tries to exploit—she's very nearsighted. Therefore his

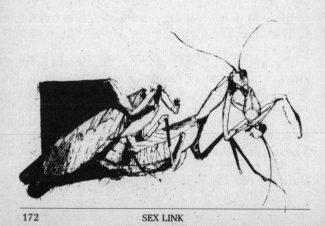

sexual approach is stealthy and could be described as attempted "surreptitious entry."

The odds on a successful performance and a getaway in one piece are heavily weighted against the male. Frequently, at his approach, the female suddenly turns at the last second and bites off his head. His sex drive, however, is so strong that even though his top half is missing, his bottom half remains inflamed with desire and mounts to copulate with her while she happily munches on his head.

After completing coitus, the decapitated male slumps to the ground, dead. She then consumes the rest of him and only then does she consider the act consummated. Well, he's not the first male to have lost his head over a *femme fatale*.

The female mantis of some species is more patient and permits the male to begin copulation before turning and snapping off her lover's head. Her next mouthful is his shoulders, followed by his arms while the unfortunate and abbreviated remainder of the male goes right on copulating.

As brutal and murderous as this practice seems, there's more to it than mere cannibalism. It's been determined that the brain of the male mantis (yes, he has a brain though it doesn't appear to be anything to brag about) tends to inhibit copulation; this procreative activity is controlled by a nervous center in an abdominal segment of the insect. Thus, when the female beheads her mate, she's actually aiding him in the performance of the sex act.

THE SEA URCHIN
Generation Gap

There's nothing spectacular about the love life of the sea urchin, but the urchin does have a highly unusual procedure of breeding only in alternate generations and a rather ghastly method of exterminating the succeeding offspring or in-between generation.

A descendant from a breeding group hatches from an egg, grows into a bowl-shaped creature, and spends its brief adult life swimming around, never reaching sexual maturity. During this aimless though pleasant period the sea urchin becomes host-mother to another organism which invades her and takes over her interior. After losing her digestive system, the mother dies and falls away, leaving a new individual in possession of her vital organs.

This young sea urchin grows and develops at sexual maturity into a creature like its grandparents. Eventually this urchin breeds and produces the intergenerational, nonsexual host-mother which is fated to become the next victim of matricide. And we complain about our generation gap!

13
POPULATION CONTROL

The great irony of our times is that if the human race becomes extinct it will be due, at least partially, to the very device invented by nature to insure the survival of our species—sex and procreation.

The problem is not that the device doesn't work; it works only too well, and it's improving all the time. Where initially it took about a million years to accumulate the first billion humans, we are now banging out that number in about ten years. It's not that we've perfected our technique because we're still using the same old method of (re)production.

The reason for our "success" lies with increased preventive measures against and control over disease. Chemical agents, higher standards of sanitation and inoculations have managed to practically wipe out the former devastating epidemics of typhoid fever, yellow fever, malaria, sleeping sickness and others. Another factor is the fairly recent introduction of sulfa drugs, penicillin, antibiotics and other miracle nostrums which have resulted in a dramatic increase of the survival rates over formerly incurable diseases.

The combination of prevention and cure has resulted in nearly doubling the average life-span of the earth's human population in the last century.

This lengthening of the average life expectancy is one of the very nicest things science could have done for us but now that we have it, what do we do about its grim consequence—overpopulation?

While some success in population control is being experienced in a few of the more industrial countries, world population as a whole is increasing at a terrifying rate. This global population explosion, if not decelerated, may soon threaten the survival of our civilization.

The great anomaly in this situation is that in contrast to the breeding capacity of most species, the human female is decidedly unproductive; she has an average of less than a half-dozen (or a potential maximum of about two dozen) offspring during a lifetime.

Contrast this with the common toad which produces 7,000 eggs at a laying, the herring with 50,000 eggs at a time, or the sturgeon which lays six million eggs at a spawning. Among other examples of fecundity one finds the tapeworm, which produces nearly 6 million offspring a year for thirty-five years, and the queen termite, which delivers 11 million eggs a year for fifteen years.

Aphids (plant lice) may lay claim to the title of champion in this field. They mature so fast that a baby aphid becomes a grandmother in twenty-four hours. Starting with a single female, a four-month reproduction period could produce so many aphids that their gross tonnage would be greater than the total weight of all the human beings now living in North and South America; provided, of course, that there's a complete absence of predators and other unfavorable conditions.

A sobering thought is that the aphid, which reproduces by parthenogenesis, accomplishes these procreative heroics without the assistance of a single male. (Or even a married one.)

While we doff our hats to the prodigious reproductive capacity of these aphids, the unquestioned champion is found among the one-celled protozoa. One individual of the genus *Paramecium* can produce, by division, 268 million descendants in a month, or about 9 million per day.

How is it that the human, with such puny reproduction capacity in comparison to other species, is in such danger of overpopulation? The answer is that all other animal life is held in check by one or more of the following conditions: natural predators, unfavorable environmental state, and restricted opportunities for mating; all of these are nature's methods of keeping all life on earth in balance. The human species suffers no predators, controls the environment for its own benefit, and has no physiological restrictions in its mating activities.

In compensation, we do have a time-honored and popular method for achieving zero population growth—a form of mass suicide called war. One tragic consequence of this ritual is that it never fails to weaken human society by eliminating its healthiest members—its youth.

Overpopulation is a prelude to disaster, but nature has provided most species with built-in safeguards to prevent such tragedies, including the limitations enforced by various natural enemies.

Lions maintain a population consistent with the resources of their environment, and this is controlled by the number of cubs that are able to survive and mature. The procedure used is the simplest, yet most effective, imaginable. When game is scarce, cubs are the first casualties—they're simply not fed.

In a lion pride, cubs are always fed last, after the adults have stuffed themselves. If there's only enough food for the adults, the cubs starve to death. Even in normal times fifty percent of the cubs die, leaving only the hardiest of them to survive.

The common mouse curtails its population by resorting to abortion under certain adverse conditions. The female is not promiscuous and breeds only with the same mate. If another male happens to copulate with her, she'll abort within four days. Merely seeing or smelling a strange male is sufficient to cause a pregnant mouse to abort.

The females of other rodents and rabbits possess this ability to abort. In normal years pregnancies are carried to full term and the population of these groups is multiplied. Under hostile environmental conditions such as overcrowding, shortage of food or long spells of adverse weather, the embryos in the female uterus do not develop. Nature provides the ideal solution in this case—the perfect abortion.

There's an unusual theory regarding the practice of birth control by the whale. These leviathans communicate with each other in musical tones, which gives the impression that they're serenading each other. Recordings have been made of these underwater songs and they indicate a sophisticated verbal language and intelligence unsuspected till recent years.

These underwater communications could well have acted as a birth control device in the years when whales were plentiful; those days, of course, are long gone. When the breeding banks became overcrowded, the multitude of singers and songs filling the water would create such a garbled cacophony that communications were overwhelmed. This would have an unsettling effect on the herd and temporarily incapacitate the whale's reproductive system.

The wolf has developed a sensible and efficient method of family planning. A wolf pack maintains a stable population geared to the food supply and the environmental conditions by regulating the number of births.

Mating couples are selected beforehand by the pack's leader and his mate who pass on the word to the prospective parents that it's their job to "get it on." Aside from this official sanction to mate, no playing around or hanky-panky is tolerated within the pack.

A sane and pragmatic method of dealing with overcrowding is also followed by the bee. When the hive reaches a certain population density, above 50,000 occupants, the bees utilize a practice known as swarming.

The queen leaves the hive, followed by a large portion of the workers, to seek out a new home in which to establish another colony. Replacing her in the hive in a complete takeover is the new queen. If the old queen were to return to the old hive inadvertently she would be summarily murdered by the new queen and her workers.

That's not exactly a heartfelt tribute to a grand old queen who's been in constant labor for several years, laying 1,200 eggs a day or nearly half a million eggs a year. We may not approve of the bees' morals or manners but they do know how to keep a place from getting overcrowded.

There are numerous groups of animals in which overbreeding rarely occurs because the periods of fertility are so distantly spaced, usually once a year. Among these are animals whose estrus periods are also extremely abbreviated, covering only a few hours in an entire year. This minimum exposure to propagation coupled with the fact that many of them deliver few offspring is nature's insurance against overpopulation for these animals. Included among many others of these low-birthrate families are the sheep, porcupine and penguin.

THE MACAQUE
Natural Chastity Belt

One group of macaque monkeys has evolved an unusual genital structure for both male and female that prevents their interbreeding with closely related species.

The female has a growth across her vagina that does not permit her male cousins to penetrate her. Her own mate, however, has a penis that's twice as long as that of his relatives but much slimmer and tapered so that he has no problem in copulating with her.

THE LEMMING
Here Comes the Mob

Aside from the human, only two other species on earth employ mass suicide as a population control, the lemming and the snowshoe hare; the former, a small Scandinavian rodent, is more widely known for its obsessive marches to destruction.

An army of lemmings remains on its course to the sea regardless of the number or the size of the obstacles confronting it. Its relentless migration ends only when the entire multitude cascades over a mountain cliff or when it reaches the ocean into which it plunges and dies.

These periodic migrations have been observed every half-dozen years or so. While the single-mindedness of a vast army relentlessly marching toward oblivion seems eerie and unexplainable, there's little doubt that overpopulation is the chief causal agent to trigger the one-way death march.

When the press of population reaches a certain density the lemming social structure disintegrates. The fact that the marches always end in mass suicide indicates that the conditions of overcrowding have such traumatic and devastating effects they overcome even the primal instinct for survival. The lemmings' choice of mass self-destruction has one advantage over our wars—they, at least, succeed in weeding out the less-fit members of their society.

THE LOCUST
Look Who's Coming to Dinner

A group of insects, lemminglike in its driven but aimless migration, is the Acrididae, the family of locusts and grasshoppers. Overcrowded conditions lead to frustration and nervousness among the insects, building up pressure that finally erupts into a spontaneous mass exodus from the environment.

A migratory locust swarm appears suddenly, without warning, and is so enormous that it darkens the sky for miles in every direction. One of the largest swarms ever recorded occurred toward the end of the last century over the Red Sea. It was estimated to be over 2,000 square miles in size. When a locust horde settles down it ravenously attacks and devours every living plant in the area while the farmer can do little more than stand by helplessly, with mixed feelings of rage and resignation.

Like the march of the lemmings, the swarming of the locusts eventually is self-destructive. But there is one great distinction between the two: before the locusts disappear they insure the next generation by mating and depositing the fertile eggs in the ground.

For the next seventeen years this generation is not visible as it spends those years as underground grubs. The day arrives when they metamorphose above ground into locusts and grasshoppers, existing peacefully in the normal environment, but if conditions lead to overcrowding, then get out of their way!

THE RAT
Murder, Rape and Anarchy

Various studies have shown the destructive effects of overpopulation on a society. A typical case study is that of a well-ordered, peaceful community composed of a number of nuclear families in a patriarchal society which also includes extensive communal interrelationships.

The authority of the dominant father in each household is unquestioned and these males also make up the community leadership. The adolescents remain at home until reaching maturity, at which time they leave to establish their own households and families.

Under these conditions of peace and harmony the population increases and the resultant overcrowding creates social disruptions. At first these are minor incidents, but they grow increasingly serious as the situation deteriorates.

The first major indication of an important breakdown of the social structure is the appearance of a generation gap, when the adolescents repudiate adult authority and form into youth gangs. These roam the community in defiance of hitherto accepted local laws and customs. They prey on all individuals they encounter and appropriate or destroy any goods or property they desire.

Soon, the spiraling crime wave with its looting, violence, rape and murder has broken down all law and order, and the erstwhile leaders, the adult males, do not dare even to leave their homes. In a short time, even the sedate housewives cannot resist the lawless excitement swirling about the community and they, too, leave home to join the anarchic scene. The bloody violence continues to mount until the population is decimated and the gangster element with its lawless fever has managed to destroy itself. Only then do the older males again take up the authority in the destroyed community and begin building anew.

This is not a description of some metropolitan inner city but the case history of a community of rats. The brown rat, with its almost human ability to learn, remember and pass on the knowledge to the next generation, is the greatest biological competitor to the human species. Yet this very intelligent animal is rendered helpless in the face of an overcrowded environment.

THE ELEPHANT
Call Me in About Five Years, Honey

Probably the most elemental technique for population control is that practiced by the elephant. The female, who takes on all responsibility for birth control, has a gestation period of twenty-two months, longest of all mammals including that of the largest whale. She has no interest in sex during this period or during the following three years which she devotes to her calf. Therefore, she mates no oftener than twice in a decade.

Elephant courtship is intense and surprisingly humanlike in its manner. After choosing a mate, the female is very coy and playful, running away from the male's advances, but not too fast nor too far. Their engagement is signaled by gentle strokings of each other's back. This is followed by a tender, lingering kiss, accomplished by inserting their trunks into each other's mouth and then lovingly intertwining their trunks above their heads.

This mutual display of affection is the beginning of a lengthy courtship during which the elephants are inseparable. During this period their feeding, resting, traveling and sleeping are mutual activities; all the while they display an affection and tenderness toward each other that would thrill the heart of any romantic.

As their courtship progresses, they become increasingly intimate in their caresses as their eight-foot trunks, each equipped with two sensitive fingers at the end, explore, stroke and titillate each other's body, including the genitals which receive special attention. Though this erotic play, particularly the mutual masturbation, is understandably very arousing sexually, they show great restraint and do not go all the way.

Finally, the female's internal clock tells her that the day for mating has arrived. For this climactic moment the elephants, which are very shy and discreet in their lovemaking, disappear into the bush where they can be alone and away from curious eyes.

Once ensconced within their honeymoon glen, the male gently mounts, lightly resting his front

legs on her back. Then everything seems to stop; there is little movement or thrusting of haunches on the part of the bull. He remains in his mounted position, his genitals joined with hers, as though resting. However, there's plenty going on. He doesn't have to move because his penis is motile. Once inserted, it takes on a life of its own, thrusting and retreating until the climax.

It's an amazingly quiet and gentle sexual proceeding, especially considering that up to 24,000 pounds of beef-on-the-hoof is joined in the venture. Nature once again showed her infinite wisdom in placing the motility in the penis because 12,000 pounds of violently thrusting bull elephant would be enough to collapse even another elephant.

THE SPRINGTAIL
Sex Among the Multitudes

Aside from microorganisms such as bacteria and protozoa, the most abundant forms of life found on earth are the arthropods (which means jointed foot in Greek), which are members of a phylum consisting of animals with segmented bodies and limbs.

The springtail is one of the tiniest members of this group and also among the most numerous; up to fifty of these minute organisms can inhabit one square inch of soil. They're found everywhere—on the ground, in the earth, around the oceans and even floating two miles high in the atmosphere. They're all around us, mostly underfoot, but most of us are not even aware of them. Their name derives from a taillike appendage which is bent and then released like a spring, enabling the insect to leap away from a predator.

Though these insects are among the very lowest orders of life, their staggering numbers tell us they're no slouches when it comes to the business of procreation. With so many different species comprising the suborder, many various methods of propagation are utilized.

One of the simplest and most common of these is also quite impersonal, if not downright boring. A male picks an area where he figures there will be females (which is just about anywhere) and then moves around, dropping his sperm willy-nilly whenever he happens to think about it. Females traveling through the area are bound to encounter the drops and become fertilized when the sperm is picked up with their vulvas.

Another method of conception is a bit more of a challenge to the male when he runs across an inviting-looking female. As she stands there, feeding or resting, he quickly draws a circle around her with drops of sperm. When she leaves the circle she can't avoid the fertile drops or the ensuing pregnancy.

A much more interesting mating game is conducted among some species where the male is much smaller than the female. He begins proceedings by clasping the female's antennae with his own and hangs on as she lifts him off the ground. For a while, he just goes along for the ride as she unconcernedly goes on about her daily business of feeding, washing and resting. This proximity eventually gets her in the mood for lovemaking whereupon he hops down and emits a drop of sperm on the ground. He then does a sort of dance with her, leading her until she's directly over the semen which she takes into her vulva.

Under favorable environmental conditions, these varied propagation methods produce a population explosion of such dimensions that the arthropods can be scooped up by the shovelful.

This situation demands swift population control measures which the insects carry out in a violent and abhorrent way.

The crowded conditions produce a tension so intolerable, these ordinarily most peaceful of creatures become frenzied and cannibalistic, with the stronger springtails attacking and consuming the weaker members. Mass migrations also take place under these and other unfavorable environmental circumstances.

THE PILL
AND OTHER CHEMICALS

Possibly the most effective population control for many animals is through the use of chemicals. No, the animals don't produce the chemicals— humans do. While we find it impossible to curb our own population explosion, we've been extremely successful in slowing down the birthrate of many animal species.

In fact, we've even caused some of them to become extinct while many other groups are on the endangered species list. We've done all this without even trying.

The best known of these birth-control chemicals is the pesticide, DDT, which produces birth defects in many newborn animals. While the use of DDT is now outlawed in many parts of the world, its residues continue to destroy animals and pollute the biosphere. DDT's successor, another chemical called PCB, is turning out to be fully as life-threatening to the animals and degrading to the environment.

Our indiscriminate use of pesticides is only part of the story. Our discharge of industrial wastes, oil spills, overintensive utilization of farmlands and other unwanted consequences of a highly industrialized world society has polluted most of the globe's bodies of water.

The proliferation of humans has resulted in the expansion of existing cities and the building of new ones, explosive increase of agricultural lands and uncalculable miles of concrete highways. This progress has reduced the living space for numerous animals and shut others out completely. Having little or no space to live in is more than enough reason for many species to curtail their propagation activities.

All of these inroads, while effectively decreasing animal populations, were unintentional and unpremeditated effects of civilization and technology. However, there remains another area where humans have consciously and deliberately moved into the field of animal birth control.

New York City, with a pigeon population of over five million, has tried various means such as poisons, traps and electric shocks in attempts to reduce the huge flocks—all to no avail. Now scientists have concocted a version of the Pill for pigeons. This is a grain the birds are fond of, which is impregnated with a chemical that inhibits ovulation in the females. While copulation among the pigeons will continue unabated, the sexual unions will be sterile, resulting in the withering of the flocks through attrition.

Switzerland, also facing a pigeon population explosion, is experimenting with a similiar technique, as is Manchester, New Hampshire. On the same subject, the U.S. Patent Office has officially approved patent No. 3,419,661 which is an oral contraceptive for birds.

A birth-control pill for dogs has been developed but is not on the market because the Food and Drug Administration fears that it might be taken inadvertently (or purposely) by humans. However, a new canned dog food containing a synthetic birth-control hormone has been tested in several cities across the nation.

Another attempt to lower the pet population by restricting the sex life of domestic animals is a local ordinance passed by the City Council of Stanfield, Oregon, a couple of years ago. This law, more social and political than scientific, prohibits the performance of sex acts by animals in public. It's doubtful that this ordinance will have any effect since there's no penalty involved for the animal criminal; instead, its owner may be fined or sentenced up to twenty-five days in jail.

SEXUAL ODORS— THE KEY TO BIRTH CONTROL

Among nature's most cunning devices for promoting procreation are hormonal substances secreted by the individual, known as pheremones or sexual odors. The female in these species releases the pheremones when she's ready for mating and the odor attracts the appropriate males which arrive ready for breeding. This accounts for all that sniffing of female genitals by the male that we observe, especially among our pets.

In most species of the order of Lepidoptera, such as moths and butterflies, the receptive female lures lovers from great distances when she releases her pheremones. Males of some species also emit sexual odors to indicate their state of fertility and readiness to mate.

The human female is included among the many species that produce pheremones. Over the course of many aeons, the practical effects of this sexual aid in humans has greatly diminished. On the one hand, the modern standards of personal cleanliness with daily bathing, perfumes, deodorants and vaginal sprays has almost completely wiped out any evidence of pheremones.

In addition, as a species we no longer have a need for an acute sense of smell as an aid to survival, and this faculty has greatly deteriorated. As a comparison, the human nose contains 5,000 sensory cells whereas the nose of a German shepherd dog has 225,000 cells. That's why a dog, along with many other animals, gets a lot more mileage with his nose than we do. Not only does he breathe through it but he also uses it to lead him to romantic liaisons.

The use of pheremones in mating requires that the organs of smell in the mates thus summoned be in good working order. Mice can even produce sexual odors that speed up or slow down each other's sexual response.

Many animals that have lost their sense of smell, through injury or disease, will not mate. This fact has not escaped the attention of scientists who have been basing many of their animal birth-control experiments on pheremones. The initial step is to isolate the female sex odor, analyze it for its chemical components, and then reproduce the precise scent synthetically. Once this female scent is captured, the scientists can proceed to utilize it in various ways.

Some scientists are working on the elimination of a major cotton pest, the bollworm. Using the female sex odor, they lure the males to traps and then sterilize them with a chemical treatment. Though sterile, the males remain sexually competitive and mate with the females in a sterile union. Once deflowered, the female loses her sexual attractiveness since the males mate only with virgins. Thus, the fertile cycle of the insects involved is permanently broken. Synthetic sex attractants are used similarly to control the gypsy moth, a major pest to forest and shade trees.

Once a pheremone has been synthetically reproduced, it can be used in various ways to deceive the males or females of any particular species to prevent mating, fertile or otherwise. Earlier in the book is a description of how a threatened invasion by the oriental fruit fly was stopped in its tracks. The flies were so confused by the use of synthetic female sex odors, they sexually attacked trees, telephone poles and street signs which had been sprayed with synthetic sex pheremones. Instead of receiving the warm embraces of passionate females the insects were engulfed by miasmal chemical residues which meant curtains for them. Deceiving insects by disrupting their mating habits certainly is far sounder ecologically than trying to destroy them with poisonous sprays that eliminate all insects, both destructive and beneficial.

14

FAMILY LIFE

A few generations ago the extended family, under the press of changing conditions, in particular the shift of population to large urban centers, reduced itself to a smaller group, the nuclear family of parents and children. In recent years this smaller unit has begun to unravel around the edges.

Dissatisfaction with marriage in general and the rising rate of divorce in particular point up how far we are from the ideal living arrangement, if there is such a phenomenon. In our search for personal bliss, various kinds of living arrangements are presently being tested, though the traditional family unit is still far and away the most popular.

Animals are much freer in their mating behavior than humans in one important respect. Except for temporarily held territories in some groups, animals are not encumbered with worldly possessions that have to be distributed when the involved parties decide to leave each other. Their mating customs are instinctual, having evolved over vast paleontological eras.

While family structure remains constant in any group, the diversity of such arrangements is unlimited among the million known species on earth. These range from loners which, except for seasonal mating breaks, live a solitary existence, to the animals that congregate in large herds, shoals, swarms and colonies.

A further classification breakdown occurs among those species where a pair mates for life (very rare); where most of the males in the group are excluded from mating and only a few of the males service all the females (also rare); the harem-type herd with one male and a bevy of females and offspring comprising a family grouping; and the arrangement followed by the great majority of animals where temporary pairing and promiscuity are the norm.

Among the monogamous animals, many of which remain paired for life, are some seemingly unlikely predatory types such as the wolf, coyote, fox and jackal. The beaver also falls into this group but since we know how hard he works to construct his home with a private pool and dam, it's sort of expected that he have a stable marriage. Incidentally, the beaver's sexual technique is quite unusual; the couple copulates while sitting and facing each other.

Several bird species pledge their troth forever, including the mockingbird. However, the female leaves her mate when the young have flown the nest. She has to move out (only to the next tree) in order to live alone so that she can sing. Through the autumn and winter the two birds sing to each other, but when spring and mating season arrive she moves back to the family nest and of course stops singing.

Other lifelong soul mates are the Australian lyrebird, greylag goose (described earlier), the raven and some species of eagle, parrot, owl and jackdaw.

Among the animals that lead solitary lives, except when the mating urge strikes, are the bear, tiger, leopard, marten, mole, hamster and oyster. The shrimp also is included in this group except for the members of the *Hymenocera* genus. This is the only invertebrate that pairs and does so for a very good reason. The female's sexual receptivity has only a five-hour span so when she needs a lover she needs him quickly.

These are the more common forms of societies found in the animal kingdom. Some of them will be described in the following pages.

THE CHIMPANZEE
Come and Get It, Fellas,
I'm in the Pink

The closest relative to the human, the chimpanzees, band together in small, sort of extended-family groups of adult males and females plus offspring of varying ages. Within the larger group are smaller versions of the human nuclear family comprised of a mother and her offspring. The males are not part of these smaller groupings since in their polyandrous culture, there's no way of knowing who the father is of any offspring.

The chimpanzees are forest dwellers, living both in the trees and on the ground, though they always berth down in the trees, making crude sleeping platforms with branches and leaves. Except·for the very young who sleep with their mothers, all other individuals sleep alone.

A group inhabits a forest area of perhaps ten miles square and moves through this territory, feeding, resting, socializing and rarely sleeping two successive nights in the same spot.

While there's a fairly free interchange between the sexes the males usually congregate together under the leadership of a dominant, or alpha, male. It's not uncommon for the leader to have a very close male buddy and in this case they assume the one-two positions in the pecking order.

The other males and females show their respect to the leader by uttering soft pant-grunts of submission, crouching and placing one hand on the alpha's groin or kissing his thigh and proceeding to groom him.

Generally, the males casually ignore the females until one of them comes in estrus, which normally occurs midway between her menstrual periods. At estrus, the skin of the female's genital area becomes swollen and usually turns pink. As though this neonlike sign were not enough to advertise her condition, she also emits a slight sexual odor from her genitals during this period.

At this time the female is not only agreeable to mating but is alert to all such interest evidenced by any and all males. One indication of male sexual excitement is when his hairs stand on end. However, this is not a sure sign since his hairs also bristle when he's nervous or threatening.

A more reliable signal is the courtship posture when the male squats, shoulders hunched and arms slightly extended. Another sexual expression is for the male to shake a branch at the watching female. At either of these invitations the female immediately moves over to the male and presents herself in the copulation posture.

In the sexual presentation, the female crouches and offers her posterior to the male who proceeds to enter her, squatting in an upright position, seemingly quite nonchalant. Copulation is brief, without the slightest attempt at foreplay, and the entire proceedings take about ten to fifteen seconds of the male's thrusting before he ejaculates.

This copulatory scene usually is played before an audience of interested spectators—most of the males in the colony. As soon as one male climaxes and withdraws he's replaced by another male and this succession of sexual partners continues until all the males have been serviced, including the adolescents who, however, get at the end of the line.

The entire proceedings—a sex encounter of one female with a dozen or so males—is accomplished with remarkably little turbulence. The female is a completely submissive receptacle and the males patiently wait their turns with no jostling to get at the head of the line.

The only commotion attached to the event usually is provided by the very young and exuberant infants who sometimes climb all over the copulating couple. The adults suffer this indignity in good humor, without irritation, and don't even push the youngsters aside.

Though the male's performance is exceedingly brief his powers of recuperation are considerable. Along with most of the male contingent, he copulates with the female several times a day. It's not surprising that at the end of the ten-day estrus period, during which the female has been subjugated to scores of sexual encounters, she's bruised, battered and worn out.

Occasionally, at the expiration of estrus, when the female's pink swelling has subsided to normal color and proportions, a curious male may not be certain that the party's over. To determine that the female is no longer sexually receptive, he inserts his finger into her vagina and then sniffs the finger. Absence of the female pheremone satisfies him that she's sexually finished, for a while.

The male youngster is much more sexually precocious than his sister. Though he doesn't reach sexual maturity until age seven or eight (the same for the female) he tries to emulate the adults when he's only a year old by approaching the sexually receptive female and attempting to mount her. Many times the female obliges by crouching down and the infant male simulates the adult thrusting motions.

When two or more females come to estrus at the same time there will be some rivalry as each tries to beat the other to the willing male. This rivalry usually becomes more intense when a mother and her young, but mature, daughter show pink swellings simultaneously. The competition may be so keen that one will even push a lover off the other in the midst of copulation and then offer herself to the male.

From our own frame of reference we might expect the male, in such a circumstance, to prefer the fresh, youthful nymphet over the aging, sexually used mother. Such is not always the case. Observations have shown that the oldest, most decrepit female in a group often is the most desirable sex object to the males.

As our closest biological relative, it's not surprising that the chimpanzee shares many physical similarities with us. The female has a menstrual cycle and ovulation and gestation periods that approximate those of the human female. There are many other physiological and sexual similarities.

There are also many differences. Of no small importance, the human seems to be the only female among the primates that experiences orgasm. One reason may be that the other primates spend no time in sexual foreplay and the sexual act itself is extremely brief, allowing little time for female arousal. Further, the human is the only female among the primates to possess fleshy, protuberant breasts and buttocks; also, full, sexually sensitive mucous lips. All of these human attributes have developed primarily as sexually significant aids in that they attract and arouse the male for the purpose of copulation.

THE GORILLA
Sexual Score: Feeble Drive,
Sophisticated Techniques

Most closely related to the chimpanzee is the gorilla, though the mature male, or silverback, at 400 pounds plus is four times as bulky and heavy as the mature male chimp. The gorilla, like the chimp, lives in small groups of several adults and offspring with about a fifty percent higher ratio of females to males.

Most of the remaining wild gorillas are found in the mountains of East Central Africa. They pursue a leisurely, nomadic life-style within a forest territory which they've staked out for themselves. Though much more terrestrial than the chimpanzees, the gorilla females and young spend the night in tree platforms. The male, because of his great weight, sleeps on the ground though he's an adept tree climber when necessary.

The gorilla is introverted. Except for the youngsters, this primate displays very little social interaction within the group. This retarded interrelationship extends to its sexual behavior as it has a comparatively weak sexual drive. Whereas the immature chimps indulge in much sexual activity such as masturbation and precocious attempts at copulation, the young gorillas evince no interest in sex until they become mature, about nine years old for males and seven years for females.

Also, where a female chimp's estrus stirs up the entire male contingent, a gorilla female's sexual receptiveness leaves the males cool. She usually has to take the initiative by approaching a male and presenting herself.

Even when a couple is engaged in copulation, the other males pay scant attention to the proceedings. Considering the huge size of the mature male he possesses a diminutive penis, averaging two and a half to three inches in length.

There is one aspect of sexuality in which the gorilla excels—its ability to adopt various positions for sexual intercourse. While the other primates and most other animals are limited to one basic position in sex, the gorilla approaches the human in its flexibility and variety of such positions.

The most common position for mating, as is true with the other primates and mammals, is to mount the female from behind. Even here, the gorilla has three different options: with the female on all fours, he stands behind her; with the female squatting, he squats behind her; with the female standing and bent over, he stands behind her.

Two other positions unique only to gorillas and humans involve copulating in a face-to-face posture. This has the female on her back with the male squatting between her legs; and the female on her back with her legs wrapped around the squatting male's back.

From these descriptions it appears that the gorilla makes up in variety what he lacks in quantity. Though his sexual frequency is much lower than that of the chimpanzee, his staying power is superior as the duration of his coitus is measured in minutes rather than seconds.

Before the commencement of copulation, the male does a strut, walking around the female with stiff legs and rigid body. The sexual act itself is very noisy as the male emits a series of loud hoots, sounding like "oh-oh-oh-oh-oh," which increase in volume and intensity as he reaches his climax.

THE ORANGUTAN
The Swingers

The third group of the great apes is the orangutan which stands between four and five feet in height and is found in the swampy, coastal forests of Borneo and Sumatra. In contrast to the mainly terrestrial habits of the gorilla and the combined terrestrial-arboreal life-style of the chimpanzee, the orangutan is almost completely arboreal, rarely leaving the trees for the ground.

Its arboreal life has given the orang extremely long and powerful arms which it uses to swing through the trees. This facility has given the ape the capability of adopting a most unorthodox sexual position—it is able to copulate while swinging from the branch of a tree or the bars of a cage.

A side note regarding our closest relatives in the animal world, the great apes, is that the three groups just described are in great danger of becoming extinct. The chimpanzee's situation is a bit less precarious than that of the gorilla. The gorilla population is dwindling as encroaching civilization reduces the forests which are its habitat. The orangutan's future is even more tenuous as its small number of survivors is contained in several shrinking forests on a couple of islands in the East Indies.

THE BABOON
It Pays to Advertise

Baboons are large monkeys with short to medium tails and long, naked, doglike muzzles. Though lower on the evolutionary scale than the anthropoid apes, they display some humanlike qualities lacking in their more advanced relatives.

For one thing, like humans, they're adaptable to almost any environment such as forests, mountains, valleys, desert and seashore as opposed to the anthropoids which could not survive without trees and forests. Also, baboons are omniverous, with a diet that includes almost everything that is edible.

In their sexual proclivities baboons far outstrip the apes in humanlike practices. Though they live in large groups and are polygamous, some species pair for brief periods. At the outset of such a liaison, sexual activities are high and the male is unusually tender and generous to his mate, even to sharing his food with her. This initial passion soon subsides, especially with the male whose libido quickly fades and is aroused again only when another female comes into his life. The female of the pairing is no more sexually faithful than the male but her indiscretions are conducted in a more furtive fashion in order to deceive her cuckolded mate.

Some female baboons have strong sex drives and solicit copulation even in the absence of estrus, which is unusual for any animal including other primates. Some even continue to desire sexual intercourse after becoming pregnant which is even more unusual in subhuman primates.

The female's sexual excitement is at its highest when she's in heat. The male shares this strong sex drive. With the continual opportunities for coitus, he's quick on the trigger and climaxes within seven or eight seconds after penetration.

The baboon is much less gentle than the anthropoid apes and, as a result, a form of rape occasionally takes place. This occurs when an excited mature male tries to force a young, immature female to copulate with him. In her frightened struggle she may be injured by the stronger male.

Sexuality makes an early appearance among the young baboons, evidenced by incidents of masturbation and homosexuality. These practices mostly disappear after the young males reach maturity and have access to receptive females.

The mandrill, a large West African baboon, once seen is not easily forgotten. To begin with, he possesses a colorful set of genitalia—a crimson-colored penis flanked by blue scrotal patches. As though this dazzling sexual display were not enough, he mimics the colorful arrangement on his face with a bright red nose and naked blue cheeks. His sex advertisements are not easily ignored when he's facing you. Even if he turns his back to you he still stands out in a crowd as he shows naked buttocks of bright crimson. The female is quite humdrum in her coloring and looks no different from other baboons.

Another species of baboon, the gelada or lion baboon of Ethiopia, finds the female favored with a double set of sexual signals. Bright red skin surrounds her genitals with the labia a deeper red, all bordered by small, white protuberances. A duplicate array of these colors appears on her chest with a patch of naked red skin surrounded by small white protuberances, and in the center of the patch are two deep red strips that mimic the vulval lips.

The human female has a physical arrangement that is not too dissimilar, with protuberant breasts and red lips that correlate with her genital area—buttocks and labia.

Very common social behavior among baboons involves sexual actions, for instance when a female is threatened by an angry male. At this, the female presents a sexual invitation which the male always accepts. The male's response, mounting and performing a few perfunctory thrusts, rarely fails to appease his anger.

Precisely the same sexual maneuvers are followed by two individuals of the same sex, for instance when a low-ranking male presents his rump to a superior male, or when an inferior female presents to a superior female. In all cases, satisfaction results when the dominant-submissive roles have been confirmed.

OTHER PRIMATES
Polygamy Works

In all, there are about 200 different species of primates with little conformity among them as to sexual behavior and structure of social systems. We already have some acquaintance with the lifestyles of the chimpanzee, gorilla, orangutan and baboon.

The gibbon is the fourth member of the anthropoid ape classification. It is the smallest and most arboreal of this grouping. Its social structure differs from the other three in that it practices monogamy within a nuclear family of two parents and several offspring. The gibbon takes a proprietary interest in a small territory which it guards against all intruders.

Spider monkeys and other long-tailed monkeys live in large societies which contain polygamous groups plus assorted aged individuals and immature males. The howling monkeys of South and Central America have adopted a clan system with about three adult males and eight adult females along with a half-dozen to a dozen offspring comprising the average clan in which polygamy is practiced.

It's not known which of the above societies, if any, represents the prehominid state of humankind.

Most of the primate societies feature promiscuity and polygamy in their socio-sexual structure. Apparently this arrangement provides excellent results. Sexual jealousy, so common in the human species, seems to be largely absent among subhuman primates.

Generally, the polygamous female in heat mates with any male in the group. In some species the female copulates with all the males during her estrus but begins her activities with the lowest-ranking member, the omega, and works her way up the hierarchy to the alpha. There's an advantage to this upward progression which, at first, doesn't meet the eye. Toward the end of her estrus her eggs descend into the uterus, and it's at this time of greatest fertility that she copulates with the leader. Thus, she succeeds in keeping all the males content while assuring a strong continuity of the species by being fertilized by the leader.

THE WHALE
Big Is Everything

The largest animal ever to have inhabited the earth is the whale, and the biggest of these leviathans is the blue whale. The blue whale reaches a length of 100 feet and a weight of 175 tons, which makes it about four times larger and heavier than the biggest dinosaur. Fortunately for all ocean inhabitants the blue whale is not predatory. That's probably because it has no teeth and so has to satisfy its appetite with tons of plankton and tiny shrimps at an average meal.

The king of the ocean is the sperm whale which, while not nearly as large as the blue whale, does have teeth. It also has a harem of up to thirty cows which it services once a year during breeding season.

The whale's breeding procedure should be of some interest (especially to whales) because of their huge size and watery environment. In order to procreate, the male has to insert his penis into the female's vagina, just as all other mammals do. Customarily, the couple lies on the ocean surface, side by side, to copulate. Sometimes a third whale helps out by lying against the other side of the female, keeping her steady and in position during the sexual engagement. This makes a lot of sense, as being nudged, even gently and lovingly, by 100 tons or more of blubber can certainly change your position.

As a change of pace, some whales mate in a vertical position, rising out of the water, belly to belly, standing on their tails. There's no room here for awkwardness or fumbling around in this position because the entire operation takes about eight to ten seconds. This is quite a feat when one considers the size of the organs involved. The blue whale's birth canal is commodious enough to accommodate the passage of a whale infant measuring twenty-three feet in length.

One can only imagine the size of the male's penis if it is to make any impact whatever on the enormous vagina. A reasonable estimate of the penis size would make it equivalent to the length and girth of a professional basketball player—the tallest one.

At any rate, once the bull has served his harem, he stores his penis away in a special body pouch for another year. Otherwise, if his member dangled freely from his body, it would act as a dragging anchor and really slow down his swimming and diving performances.

The newborn calf feeds on mother's milk, holding the nipple in the corner of its mouth. The nursing blue whale mother must have lots of milk because her infant gains weight at the rate of 200 pounds per day.

THE SEAL
No Eating or Sleeping but
Lots of Loving and Fighting

The Northern fur seals roam the ocean from North America to Japan during the fall and winter migratory season. During this phase they are very social, mingling freely with seals of other species. When mating season arrives, they all head back to their birthplaces. The bull arrives at the rookery on one of the northern islands in early May, about a month ahead of the females, to select a strip of territory on the beach and then defend it against all male intruders.

At this time the seal's personality undergoes a profound change; he is transformed from a peaceful, friendly animal to one that's aggressive and mean-tempered, concerned solely with jealously guarding his proprietary rights. When the females arrive in early June, these aggressive attitudes intensify. It's every bull for himself as they begin to collect their harems. The air is full of threats and counterthreats mixed with numerous violent physical combats over the possession of the mating territories and the females.

Even after collecting his harem, the male's problems are far from over; his job now is to defend his possessions from being usurped by the cocky young bulls that still come along to challenge him. Also, he must keep an alert eye on his flock of females to keep any from sneaking off to some other bull.

The females give birth shortly after arriving at the island, usually within 48 hours, the result of having conceived the year before. After giving birth the female must wait a week before she can be mated and this is the ticklish period. If the bull relaxes his guard a cow may slip away for an illicit mating and he will have lost his fiercely fought-for privilege. Therefore, the proprietor doesn't sleep during this supercharged period and doesn't even eat; in fact, he hasn't eaten since his arrival at the rookery, more than a month earlier. His top priority at this time is to defend his territory and to keep the females in place until he's mated with them. Until she's been serviced the female is not allowed to go into the sea for food or a swim.

A harem can number from just a couple of females to a hundred, with the average being about thirty. Of course, the larger the harem, the more difficult it is to keep it intact. The larger harems are ruled over by the larger and more dominant bulls. This system of sexual selection warrants that the offspring will be sired by the largest, strongest and most aggressive bulls in the herd, thus continuing and even improving the stock of the species.

THE BEE
Fifty Thousand Virgins

One of the most disciplined and orderly societies in nature is that of the bee. The colony has a population of about 50,000 bees, practically all of them female workers such as nurses, builders and food gatherers. The few drones are all males and their only function is to be available when a new queen has to be mated, so they have very little work to do. That's because the queen bee, once mated, remains fertilized for the remainder of her life—about five years. During this time, she lays about 1,200 eggs a day and makes decisions whether the eggs are to be male or female, or sterile or fertile. Sterile females are the workers, fertile females are future queens, and fertile males are the drones or loverboys.

In a rare emergency, when there is no queen in the colony, the workers are able to create one. They have a glandular secretion known as royal jelly which they feed to the sterile worker larvae. Its hormone content is so high that it causes otherwise sterile eggs to hatch into fertile females, one of which becomes the new queen.

This kind of sex-change operation is vital to the survival of the colony and must be done swiftly. Since the workers have a forty-two day life-span, a hive without a queen would be quickly decimated and would suffer complete extermination at the expiration of those six weeks.

The bee is endowed with incredible eyesight and smelling power. Its two large eyes consist of 12,600 separate, long-focus telescopic lenses, placed at slightly different angles so that the bee can see tiny objects and slight movements at great distances. This compares with the human eye which has a single wide angle lens. The bee has 4,800 smelling discs in its antennae so what it doesn't see, it smells.

It also has a fantastic navigational sense with the ability to fly to a pollen-filled flower several miles away and then return directly to its starting point. Apparently, it uses the shadows cast by the sun as its lodestone.

Even more amazing is its ability to transmit information about the pollen to its fellow workers. Upon returning to the hive a bee scout does a little dance on the interior wall. The angle of its body gives the direction of the flower pollen and its height off the floor gives the distance. These directions are so precise that a swarm of bees will set out and fly directly to the pollen-laden flower.

However efficient and dedicated the bee is as a worker, its sex life is a dismal zero. Out of a 50,000 hive population only two individuals experience the act of sex, the queen bee and a drone and that's only on a single occasion. This occurs on the nuptial flight of the virgin queen. At this royal event, the queen is extremely fussy about the weather and will postpone the honeymoon until the day is sunny, windless and cloudless. On such a weather-perfect afternoon she'll zoom out of the hive with about 200 eager males hot on her tail.

These studs, incidentally, are magnificent specimens, specially bred for the mating occasion and possessed of size, strength, eyesight and smell that is far superior to the ordinary bee. Despite these endowments and the fact that this is the female's maiden flight, she's able to easily outmaneuver the male contingent and frustrate all their sexual attempts, until she's ready. At the very height of her flight she slows down imperceptibly and one of the studs swoops upon her to make the penetration. They continue flying and copulating in tandem and it's all over in a couple of seconds.

When the male separates, his penis breaks off and remains in the queen's vagina as a plug to prevent any loss of sperm. The male then falls to the ground where he bleeds to death. That appears to be a harsh ending for the winner, but the fate of the 200 losers is even more harrowing. Having completely failed in their sexual objective they now return to the hive where they're deliberately starved to death.

Why the mating of a single queen bee should require the destruction of 200 magnificently developed males is one of nature's enigmas and an example of her seemingly profligate expenditure of resources. However, the balance is all in nature's favor. From that single airborne fertilization the queen will deliver, in the following five years, over two million eggs which will become, for the most part, female workers. While all this represents very little sex play it produces a lot of honey.

THE TERMITE
Nothing Is Too Good for Royalty

As far as termite sexual experiences are concerned, they're no better off than the bees. In fact, they're even worse off statistically. Where a beehive population is counted in the thousands, a termite colony numbers in the millions. In all this vast horde, only two individuals have wings and sex—the royal couple.

While the millions of termite workers and soldiers are sterile, the king and queen are very, very fertile. They're also very busy being fertile because all those millions of termites in the colony are brothers and sisters bred by one set of hardworking and prolific parents.

In addition to the sex, there are other rewards for the king and queen. A large room in the termitary, the royal chamber, is set aside for their exclusive use. They have hundreds of servants that swarm around them, taking care of all their needs—feeding, washing, and hauling away the never-ending stream of eggs. Surrounding all this activity is a small army of soldiers for their protection.

In some species the queen lays an egg every two seconds, all day long and far into the night. This adds up to a neat 11 million progeny a year and goes on for twelve years or more. While the queen is delivering all those eggs, the king sits beside her except when he's busy fertilizing her, which is quite often.

After à while, this intensive breeding starts to take its toll on the queen. She loses her wings and she starts to put on weight, and she keeps right on growing until her body looks like a blimp. In fact, it's hardly a body anymore; it's a reproduction factory with an assembly line working overtime. Even though she's no longer the trim maiden he married, the king remains loyal and stays with her. Now sex is a chore with a lot more work and a lot less romance. Every time he has to impregnate her, he must squirm and slide and push his way under her gross sausagelike body before he can reach her genitals and copulate with her. It's just another example of *noblesse oblige*.

THE CUCKOO BIRD
Maybe She's Not So Cuckoo

The female cuckoo bird has adopted a life-style in which she repudiates the widely accepted aphorism, "There are no free lunches." She manages to sail through life unburdened with the usual nitty-gritty of domesticity and maternal duties.

Her procedure is ingenious in its utter simplicity. She just lays her egg in some other bird's nest and then takes off, leaving the entire regimen of brooding, hatching, feeding, safeguarding and training to the unwitting host. The cuckoo bird does select its nest rather carefully, taking pains to pick proxy parents that produce eggs similar in size and color to hers. Otherwise, her egg, if too different, might be destroyed by the host bird.

Once hatched, the cuckoo fledgling employs an instinctive backward shoving motion that frequently succeeds in pushing everything out of the nest, including the eggs and offspring of the host. The result of all this is that the young cuckoo bird becomes the sole survivor and benefits by the attendant superior care and attention of its foster parents. In some cases, the cuckoo infant develops to a size that is larger than that of the hosts. If the difference between them is great enough, the parents have to stand on the baby's head in order to feed it. That is the ultimate indignity.

In the meantime, the mother cuckoo, with nothing but time on her hands and mating on her mind, drifts around, copulating with any available male of its species. Following its parasitic style, it may have up to twenty such affairs in a mating season, depositing and abandoning the twenty resultant eggs in the nests of twenty unsuspecting and trusting foster parents.

THE HORNBILL
What Confinement Really Means

One nest that never could be invaded by the cuckoo bird, or by any other bird, is that of the hornbill, a large-beaked bird of Southeast Asia. During a lengthy period of confinement, its eggs are hatched and the fledglings allowed to mature in tomblike security.

When egg-laying time approaches, the hornbill

pair finds a suitable hollow in a tree which they proceed to close off. Working like a pair of masons they construct the wall; he, on the outside, uses mud and resin for his plaster and she, on the inside, is limited to using her excrement. When completed, the nest is sealed from the outside world by a concretelike wall.

A single feeding hole, large enough only for her beak to protrude, is the only opening to the hermitage. For the next two months the male works his tail off, hunting and delivering food to his walled-off family. This could amount to several hundred feeding trips a day.

When the young chicks are ready to leave the nest, the mother breaks the wall down with repeated blows of her powerful beak. Undoubtedly, the most relieved member of the family at this time is the father who will now have the time and luxury of feeding himself.

THE TALEGALLAS
Commune Living

In the mating season, many species of birds congregate in large flocks to pair off and build their individual nests. A New Guinea bird, the talegallas, goes far beyond such activity in the integration of a small society.

A group of these cocks cooperates in the construction of a large communal nest for the breeding area of a number of bird pairs. The structure rises to ten feet in height and sixty feet in circumference. The interior is apportioned among the males, who build individual nesting chambers for their mates to brood in. This is probably the largest structure built for nesting by any vertebrate—aside from the human.

THE FINCH
The Home Wreckers

Some birds aren't limited to an annual mating season but mate throughout the year. This breeding flexibility is found among some species of the large songbird family of finches. Because of this ongoing courting and mating, these birds form fairly permanent pair-bonds, for convenience, if nothing else.

These relationships can be fairly fragile at times, evidenced when a new female appears on the scene. At such time the natural male reaction is to welcome the newcomer by trying to rape her. Unless the female cooperates fully, the rape attempt becomes just another exercise in futility.

If the rape doesn't work out, the male may try normal courting procedures. Frequently, his mate doesn't take too kindly to the interloper and she fights hard to get her mate back to the nest. She follows him around and goes through the entire set of courtship rituals of dancing, hopping, bowing, feather ruffling and sexual crouching repeatedly, over a long period of time. Interspersed with her courtship endeavors are physical attacks upon the intruding female who usually doesn't fight back.

The male eventually must make a decision. If it's to return to his mate, they usually combine to attack and drive away the new female. If it's the new bird that grabs his fancy, he'll become more intimate with her by copulating and building her a nest. With love and a home to back her up, the newcomer becomes more confident and begins to fight and even attack the old mate. At this point, the male often joins in and the two of them succeed in driving off his old mate.

The same procedures occur if the new member of the triangle is a male.

THE CRICKET
Even the Lonely Need Sex

The cricket is a loner, and the male spends most of his solitary life in a small crack or furrow. His only opportunity to get into a relationship is to attract a female for mating purposes. That's why we hear a cricket chirping (he does this by rubbing his wings together) sometimes for hours. It's his attempt to break up his lonely existence and get some sex at the same time.

Since the male doesn't get around very much, it's a good thing that the female is more peripatetic. When in the mood for company and sex (after all, she lives alone too) she seeks out a male, being guided by his chirps.

Once the two lonely and sexually aroused insects meet, she strokes him with her antennae and his loud, clear mating call changes to a softer courtship song; at the same time he backs toward her with a swaying motion. The female climbs up on his back where she feeds on a tasty glandular secretion while he tenderly strokes her with his antennae.

When she's gotten enough of the glandular goody and has climbed far enough up on his back, they're ready for copulation. She lowers the rear tip of her abdomen as he raises the tip of his abdomen and when the two meet, his genitals insert a sperm packet into her vagina.

Having experienced a gratifying sexual encounter, the male doesn't want it to end there and he breaks into another song which could be titled, "Stick Around Baby." This love song plus the love nectar on his back would be enough to keep her there because she also digs the action; but there's yet another attraction for her. That's his sturdy front wings on which she loves to nibble while copulating. She stays on and they copulate again, and again, until both are content and surfeited. Except for occasional rivals that he has to chase away, it's a loving and peaceful interlude for two otherwise lonely souls.

THE KANGAROO
Room Service in a Pouch

The kangaroo is a marsupial which is one of the classifications of mammals. Though these marsupials form one of the lowest orders of mammals, this is the first viviparous (live birth) mammal. The main difference between marsupials and higher mammals is that the marsupial embryo goes through a very brief period of incubation, only two or three weeks and, while born live, is still undeveloped.

The newborn kangaroo is entirely naked and blind, with ears barely visible and hind legs weak and undeveloped; it is completely helpless, and about an inch in length. The full-grown kangaroo is eight feet in height which underscores the diminutive size of the new infant.

At birth, the tiny embryo has to make its way from the mother's birth canal to the stomach pouch which contains the teats to the milk glands—an epic journey for such a helpless organism.

The mother prepares for the impending event by cleaning out her fur-lined pouch with her paws and tongue. Sitting with her tail between her hind legs, she makes a groove along the center of the tail with her tongue, from her cloaca to the pouch.

At parturition, the infant drops on the tail and follows the track, using its clawed forelimbs to propel itself, until it reaches the pouch and falls in. The embryo moves around the pouch, its mouth wide open, until it finds a teat to which it fastens its mouth. Once joined to the life-giving teat, the infant doesn't let go for the next few months. In fact, it couldn't let go even if it tried because once the teat is taken into the mouth it expands so that it can't be removed.

After completing its weaning and attaining considerable growth, the young kangaroo remains in the pouch for a while, leaning out from its refuge to feed on vegetation.

COUVADE
Male Childbed

Couvade is a primitive custom that has the father go to bed when his wife prepares to give birth. The father is attended to as though he were pregnant. During the actual birth he mimics the trauma of the labor by straining and groaning in pain. The custom goes back to ancient times and was widely practiced in various societies around the world.

The origin of couvade has been traced back to the time of the transition from the matriarchal to the patriarchal system of tribal organization. It stems from the father's desire to emphasize the blood bond between himself and the child.

According to ancient American Indian folklore, the father would refrain from arduous work or dangerous pursuits at the time of his child's birth and he would be extremely careful of his diet at that time. This was based on the belief that the intimate relationship was such that the activities of the father would directly affect the infant.

It comes as no surprise that within nature we find many forms of couvade among animal groups in which the male takes over many of the female's traditional duties in the care of newly born offspring. This role reversal is quite common among some species of birds and fish where the male takes charge of the fertilized eggs, guards them while they're hatching, and feeds the hatchlings if necessary.

The oyster is one group whose members don't change their sex roles—they change their sex! In some species they change their sex frequently, one week releasing eggs, the next week emitting sperm. Other types of oysters are more leisurely about their sex changes, making them annually. At any rate, no one can tell the sex of an oyster from its appearance, even on the half-shell.

A particularly dedicated and conscientious father is found among the West African black-chinned mouth brooders. After the eggs have been laid and fertilized, the male takes them in his mouth and carries them until they're hatched. This covers a period of 23 days during which time the proud father does not eat. He doesn't dare to.

THE VAQUERO FROG
The Frog Has a Frog in His Throat

In South America there's a species of tiny frogs called vaquero. The pregnant female of this species requires the services of several males to fertilize the several eggs she lays.

After fertilization, the males stand guard until there's movement within the eggs. The eggs are then swallowed by the males where they repose in a special accouchement pouch within the body of each frog. Here the embryos have food, warmth and security until ready for birth, at which time the males regurgitate the tiny tadpoles into the water.

This unusual reproduction scenario encompasses couvade, group and oral sex, all of which may be sex aberrations to some, but not to the vaquero.

THE TOAD
The Male Midwife

There's a species of European toad which not only takes over the responsibility of incubating the fertilized eggs but also acts the part of *accoucheuse* or midwife. When the female is ready for delivery, the male grips an end of her string of eggs with his lips and pulls them out of her cloaca.

He then expertly wraps the egg chain around one of his hind legs. He's not content with just one family so he services three or four females in this way until both hind legs are covered with strings of eggs and resemble a pair of galligaskins.

For a while he's extremely careful in his movements—almost like walking on eggs—and spends much of his time in wet grasses to keep the eggs moist. When the time for hatching arrives he releases the eggs into a pond or stream where the tadpoles break through their covering and swim away.

THE SEA HORSE
Daddy's Pregnant Again

In Greek and Roman mythology, one of the fabulous inventions is a surrealistic creature, half horse and half fish, driven by sea gods or ridden by sea nymphs. This, like most ancient myths, seems to be woven from pure fantasy, but this one may have some basis in fact. In the warm waters of our southern latitudes is a strange looking creature of the genus *Hippocampus* that is commonly known as the sea horse because it looks as if it's half horse and half fish.

Its appearance, bizarre though it may be, doesn't hold a candle to its sex practices. Watching the courtship of a pair of sea horses one is certain that he's observing a complete and total sex role reversal.

The female is much more flamboyant in her coloring (quite the opposite of most species of fish that change colors during courtship), and she's definitely the aggressor in the mating ceremonies.

As the two come together in the sexual embrace, she holds him close with her tail wrapped around his body. At the climactic moment she inserts her sex organ repeatedly into the middle of the male's body. For all the world it appears that she's thrusting her penis into the male's sexual opening, and releasing her sperm on his eggs. In actuality, her member is not a penis but a genital papilla (a nipplelike projection) through which she passes her eggs into a special pouch in the male's stomach. Her repeated insertions excite the male and he discharges sperm over the eggs.

Now the sex role reversal becomes complete. The male carries the fertilized eggs in his stomach pouch and the female swims away without a maternal worry in the world. The fish embryos, snug in the "womb," feed on nutrients supplied to them by the father. As they grow and hatch, his belly becomes swollen, giving him a true pregnant look.

Finally, when the baby sea horses are ready for birth, the proud and relieved father opens the sphincter of his stomach pouch and delivers the tiny brood into the sea.

INDEX OF
ANIMAL LIFE

African Blood Fluke, 34
African Clawed Frog, 79-80

Baboon, 201-204
Bat, 54
Bedbug, 39-40
Bedtick, 41
Bee, 208-211
Beetle, 134-135
Bowerbird, 140-143
Bumblebee Eelworm, 34-35
Butterfly, 154-155

Callicebus Monkey, 45-46
Carp Parasite Worm, 35
Chimpanzee, 195-199
Cichild, 127-128
Cockroach, 103-106
Cricket, 216-217
Crocodile, 61-63
Cuckoo Bird, 212-213

Deep-Sea Angler, 37
Dog, 121
Dragonfly, 118-120
Drakes, 63-65
Ducks, 63-65
Dysticid, 82-83

Earthworm, 31-33
Eel, 163-164
Elephant, 183-186

Fiddler Crab, 156-157
Finch, 215
Firefly, 54-56, 164-166
Fly, 92-100. See also Firefly,
    Dragonfly, Mayfly
Frog, 65-69. See also African
    Clawed Frog, River
    Bullhead, Toad, Vaquero
    Frog

Goose, 150-153
Gorilla, 199-201
Guinea Pig, 37-39

Hornbill, 213-214
Hummingbird, 153-154

Kangaroo, 217-218
Kob, 106-109

Labyrinth Fish, 80-81
Lemming, 180-181
Lion, 88-92
Locust, 181-182

Macaque, 180
Marine Bristle Worm, 36-37
Mayfly, 51-52
Mink, 60-61
Mosquito, 52-54
Moth Mite, 69-70
Mud Snail, 48

Octopus, 112-115
Orangutan, 201

Paramecium, 50-51
Parasitic Wasp, 70-71
Penguin, 146-148
Pigeon, 148-149
Pheasant, 143-144
Porcupine, 130-133
Praying Mantis, 171-173

Rat, 182-183
Raven, 149-150
River Bullhead, 81-82
River Crayfish, 135
Robin, 136
Roebuck, 133
Ruff, 144-146

Sage Grouse, 101-103
Salmon, 136-137
Scorpion, 169-171

Sea Horse, 221-223
Seal, 207-208
Sea Urchin, 173-174
Slipper Snail, 49-50
Spider, 115-118, 166-169
Springtail, 186-188
Snail, 75-78. See also Mud
    Snail, Slipper Snail, Water
    Snail
Starworm, 35-36
Stone Grouse, 109-110

Tallegallas, 214
Ten-Spined Stickleback, 128-
    130
Termite, 211-212
Threadworm, 33
Toad, 221
Turtle, 157-159

Vaquero Frog, 220-221

Water Snail, 46-48
Whale, 205-207
Worm, See Bumblebee
    Eelworm, Carp Parasite
    Worm, Earthworm,
    Marine Bristle Worm,
    Threadworm

ABOUT THE AUTHOR

HY FREEDMAN has been a writer for Groucho Marx, the "NBC Comedy Hour," "Bachelor Father," Red Skelton and a host of others. He has also been president of the television branch of the Writers Guild of America West. His serious hobby, for as many years as he has been a writer, has been collecting animal sexual lore. Mr. Freedman is also the author of *Supermarriage—Supersex*.

DON'T MISS
THESE CURRENT
Bantam Bestsellers

☐	11708	**JAWS 2** Hank Searls	$2.25
☐	11150	**THE BOOK OF LISTS** Wallechinsky & Wallace	$2.50
☐	11001	**DR. ATKINS DIET REVOLUTION**	$2.25
☐	11161	**CHANGING** Liv Ullmann	$2.25
☐	10970	**HOW TO SPEAK SOUTHERN** Mitchell & Rawls	$1.25
☐	10077	**TRINITY** Leon Uris	$2.75
☐	12250	**ALL CREATURES GREAT AND SMALL** James Herriot	$2.50
☐	12256	**ALL THINGS BRIGHT AND BEAUTIFUL** James Herriot	$2.50
☐	11770	**ONCE IS NOT ENOUGH** Jacqueline Susann	$2.25
☐	11699	**THE LAST CHANCE DIET** Dr. Robert Linn	$2.25
☐	10150	**FUTURE SHOCK** Alvin Toffler	$2.25
☐	12196	**PASSAGES** Gail Sheehy	$2.75
☐	11255	**THE GUINNESS BOOK OF WORLD RECORDS 16th Ed.** The McWhirters	$2.25
☐	12220	**LIFE AFTER LIFE** Raymond Moody, Jr.	$2.25
☐	11917	**LINDA GOODMAN'S SUN SIGNS**	$2.50
☐	10310	**ZEN AND THE ART OF MOTORCYCLE MAINTENANCE** Pirsig	$2.50
☐	10888	**RAISE THE TITANIC!** Clive Cussler	$2.25
☐	2491	**ASPEN** Burt Hirschfeld	$1.95
☐	2222	**HELTER SKELTER** Vincent Bugliosi	$1.95

Buy them at your local bookstore or use this handy coupon for ordering:

Bantam Books, Inc., Dept. FB, 414 East Golf Road, Des Plaines, Ill. 60016

Please send me the books I have checked above. I am enclosing $_____
(please add 50¢ to cover postage and handling). Send check or money order
—no cash or C.O.D.'s please.

Mr/Mrs/Miss_____

Address_____

City_____State/Zip_____

FB—8/78

Please allow four weeks for delivery. This offer expires 2/79.

Bantam Book Catalog

Here's your up-to-the-minute listing of every book currently available from Bantam.

This easy-to-use catalog is divided into categories and contains over 1400 titles by your favorite authors.

So don't delay—take advantage of this special opportunity to increase your reading pleasure.

Just send us your name and address and 25¢ (to help defray postage and handling costs).